Progress in Nonlinear Differential Equations and Their Applications

Volume 73

Editor

Haim Brezis
Université Pierre et Marie Curie
Paris
and
Rutgers University
New Brunswick, N.J.

Patrizia Pucci
James Serrin

The Maximum Principle

Birkhäuser
Basel · Boston · Berlin

Authors:

Patrizia Pucci
Dipartimento di Matematica e Informatica
Università degli Studi di Perugia
Via Vanvitelli 1
06123 Perugia
Italy
pucci@dipmat.unipg.it

James Serrin
University of Minnesota
Department of Mathematics
Minneapolis, MN 55455
USA
e-mail: serrin@math.umn.edu

2000 Mathematics Subject Classification 35J15, 35J60, 35J70, 35A05, 35B05, 35B50, 35R45, 58J70.

Library of Congress Control Number: 2007929013

Bibliographic information published by Die Deutsche Bibliothek
Die Deutsche Bibliothek lists this publication in the Deutsche Nationalbibliografie;
detailed bibliographic data is available in the Internet at <http://dnb.ddb.de>.

ISBN 978-3-7643-8144-8 Birkhäuser Verlag AG, Basel · Boston · Berlin

© 2007 Birkhäuser Verlag AG
Basel · Boston · Berlin
P.O. Box 133, CH-4010 Basel, Switzerland
Part of Springer Science+Business Media
Printed on acid-free paper produced from chlorine-free pulp. TCF∞
Printed in Germany

ISBN 978-3-7643-8144-8   e-ISBN 978-3-7643-8145-5

9 8 7 6 5 4 3 2 1   www.birkhauser.ch

Contents

Preface

In the years 1948 to 1950, one of us (J.S.) had the unique opportunity of attending lecture courses on elliptic differential equations given by Professors Eberhard Hopf and David Gilbarg at Indiana University. These exemplary lectures first awakened his interest in this theory and in particular in the subject of the maximum principle. The other of us (P.P.) began the study of partial differential equations at the Universities of Perugia and of Michigan with Professor Lamberto Cesari, who instilled in her a deep regard for clarity and rigor, as well as for the importance of dealing with concrete problems. This combination is the background of the present work.

The maximum principle enables us to obtain information about solutions of differential equations and inequalities without any explicit knowledge of the solutions themselves, and thus can be a valuable tool in scientific research. In particular, this book should prove useful not only to professional mathematicians and students primarily interested in mathematics, but also to physicists, chemists, engineers and economists. The maximum principle moreover occurs in so many places and in such varied forms that anyone learning about it becomes acquainted with the classically important partial differential equations and, at the same time, discovers the reason for their importance.

We consider classical linear and quasilinear elliptic inequalities as well as divergence structure and variational operators, with emphasis on the important topics of comparison results and tangency theorems. This work ultimately applies also to weak solutions in appropriate Sobolev spaces.

In order that the book may serve the purposes of reference and as a basis for further developments, the proofs are given in detail. This has led, at a number of points, to results either not found elsewhere, or not readily accessible. Many of the proofs and derivations, even of the standard parts of the theory, are new, along with the first book presentation of the modern compact support principle and the general theory of structured elliptic inequalities. The proofs here, though difficult, make the subject available for the first time to the general reader.

Problems are introduced in the conviction that no mastery of a mathematical subject is possible without working with it. They are designed primarily to illustrate or extend the theory, although the desirability of occasional concrete easy examples has not been ignored.

The most relevant related works are the classical monographs of Gilbarg and Trudinger [43] and the earlier work of Protter and Weinberger [76]. While both these books remain of essential importance and have been invaluable as background for the present work, neither contains an up-to-date modern treatment of the maximum principle itself.

Readers should find the work valuable not only for its detailed presentation, but also as a reference work and possible graduate text material.

We are grateful to Michel Chipot and Hans Weinberger for a number of suggested improvements in this work. We are also particularly indebted to Antonio Ambrosetti for his initial encouragement to us for writing this book.

Minneapolis, April 2007   Patrizia Pucci and
                                                  James Serrin

Acknowledgement

The first author was supported by the Italian MIUR project titled *"Metodi Variazionali ed Equazioni Differenziali non Lineari"*.

Chapter 1

Introduction and Preliminaries

1.1 Introduction

The maximum principles of Eberhard Hopf are classical and bedrock results of the theory of second order elliptic partial differential equations. They go back to the maximum principle for harmonic functions, already known to Gauss in 1839 on the basis of the mean value theorem. On the other hand, they carry forward to the maximum principles of Gilbarg, Trudinger and Serrin, and the maximum principles for singular quasilinear elliptic differential inequalities, a theory initiated particularly by Vázquez and Diaz in the 1980s, but with earlier intimations in the work of Benilan, Brezis and Crandall. The purpose of the present work is to provide a clear explanation of the various maximum principles available for second-order elliptic equations, from their beginnings in linear theory to recent work on nonlinear equations, operators and inequalities. While simple in essence, these results lend themselves to a quite remarkable number of subtle uses when combined appropriately with other notions.

The first chapter concerns tangency and comparison theorems, based to begin with on the pioneering results of Eberhard Hopf. Section 2.1 includes in particular a discussion of Hopf's nonlinear contributions, which are in fact not nearly as well known as his classical linear principle. We continue with a treatment of quasilinear equations and inequalities, with linear equations of course being an important special case. We consider both non-singular and singular cases, that is, in the latter case, equations which lose ellipticity at special values of the gradient of solutions, particularly at critical points. The concern here with singular equations arises from their growing importance in variational theory and applied mathematics, as well as their from specific theoretical interest, e.g., the celebrated p-Laplace operator Δ_p.

The results of Hopf apply specifically to C^2 solutions of elliptic differential inequalities. In many cases, however, especially when the equations and inequalities in question are expressed in divergence form, as in the calculus of variations, one can expect solutions to be no more than of class C^1 or even only weakly differentiable in some Sobolev space. The solutions then must naturally be taken in a distribution sense. Correspondingly, in such cases, the study of maximum principles requires new techniques as alternatives to Hopf's approach. These methods, necessarily integral in nature, originally arose from the work of a number of mathematicians, going back as far as Tonelli, Leray and Morrey in the years 1928–1935.

Sections 2.4 and 2.5 are devoted specifically to C^1 solutions of divergence structure inequalities, allowing both singular and non-singular operators. Theorem 2.4.1 and its attendant corollaries are of special interest for their simplicity and elegance; see also the corresponding uniqueness result for the singular Dirichlet problem (2.6.2). We note also the Tangency Theorem 2.5.2 obtained from the weak Harnack inequality (Section 7.1).

Chapter 3 continues the study of divergence structure inequalities, but for more general operators for which the methods of Chapter 2 are inadequate. The principal results are:

(i) the maximum principles of Section 3.2 for homogeneous inequalities;

(ii) the "*thin set*" maximum principle in Section 3.3;

(iii) the generalization of Theorem 2.4.1 given in Theorem 3.4.1 (applying to solutions in the Sobolev space $W^{1,p}$);

(iv) Theorem 3.5.1 for weakly singular inequalities; and

(v) the interesting Theorems 3.6.1 and 3.6.5 for strongly singular inequalities.

We emphasize as well the Maximum Principles Theorems 3.7.2 and 3.7.4, and the series of uniqueness theorems in Section 3.8. These results, which extend well-known theorems of Gilbarg and Trudinger for the Dirichlet problem, see, e.g., [43], Theorems 3.8.1 and 3.8.4, appear to be new in the generality given.

Chapter 4 is a digression from the earlier emphasis on tangency, comparison and maximum principles, dealing instead with two-point boundary value problems for nonlinear ordinary differential equations. This work is preliminary to the strong maximum principles of Chapter 5, but also has ramifications in some unexpected byways. In particular, there are intimate connections with the exterior Dirichlet boundary value problem and with the existence of dead cores at infinity, see Section 4.3.

Chapter 6 is concerned with maximum principles for the complete quasilinear divergence inequality

$$\operatorname{div} \boldsymbol{A}(x, u, Du) + B(x, u, Du) \geq 0, \qquad Du = \operatorname{grad} u, \qquad (1.1.1)$$

under the general structure conditions (6.1.2); see particularly the remarkable Theorems 6.1.3–6.1.5. The proofs involve application of special test functions together with Moser iteration techniques. In view of the interest and importance of the conclusions, particularly in the theory of multiple integral variational problems, we present the proofs in careful detail. As a byproduct of this work, in Chapter 7 we consider the important issues of local boundedness and Harnack inequalities for solutions of (1.1.1), under similar structural assumptions. This work allows us as well to extend De Giorgi's famous Hölder continuity theorem to solutions of (1.1.1). The proofs of the latter results rely heavily on the celebrated John–Nirenberg inequality; for completeness we include a concise analytical proof of this result in the appendix to Chapter 7.

Chapter 5 is concerned with the Strong Maximum Principle and the Compact Support Principle for singular quasilinear differential inequalities. Since these results may be less known to the reader, and at the same time are of recent research interest, we shall pay special attention to them here. Consider in the first instance the canonical divergence structure inequality

$$\operatorname{div}\{A(|Du|)Du\} - f(u) \leq 0 \qquad (1.1.2)$$

in a domain (connected open set) Ω in $\mathbb{R}^n$, $n \geq 2$. To begin with, the following conditions on the function $A = A(s)$ and the nonlinearity $f = f(u)$ will be imposed.

(A1) $A \in C(\mathbb{R}^+)$, $\mathbb{R}^+ := (0, \infty)$;

(A2) $s \mapsto sA(s)$ *is strictly increasing in* $\mathbb{R}^+$ *and* $sA(s) \to 0$ *as* $s \to 0$;

(F1) $f \in C(\mathbb{R}_0^+)$;

(F2) $f(0) = 0$ *and* f *is non-decreasing on some interval* $(0, \delta)$, $\delta > 0$ *finite.*

Condition (A2) is a minimal requirement for ellipticity of (1.1.2), allowing moreover singular and degenerate behavior of the operator A at $s = 0$, that is at critical points $(Du = \mathbf{0})$ of u. No assumptions of differentiability are made on either A or f when dealing with the canonical model.

The operator $\mathrm{div}\{A(|Du|)Du\}$ can be called the *A-Laplace operator*, to place it in the context of well-known elliptic theory. For the Laplace operator, that is when (1.1.2) takes the classical form

$$\Delta u - f(u) \leq 0, \qquad u \geq 0,$$

we have $A(s) \equiv 1$. Similarly, for the degenerate p-Laplace operator $\mathrm{div}(|Du|^{p-2}Du)$, $p > 1$, here denoted by Δ_p, we have $A(s) = s^{p-2}$, while for the mean curvature operator one has $A(s) = 1/\sqrt{1+s^2}$. A further example is $A(s) = s^{p-2} + s^{q-2}$, $1 < p < q$, which has applications in quantum physics, see [9]. Note also that (1.1.2), when equality holds, is precisely the Euler–Lagrange equation for the variational integral

$$I[u] = \int_{\Omega} \{\mathscr{G}(|Du|) + F(u)\}dx, \qquad F(u) = \int_0^u f(s)ds, \qquad (1.1.3)$$

where $\mathscr{G}$ and A are related by $A(s) = \mathscr{G}'(s)/s$, $s > 0$. Condition (A2) implies that $s \mapsto \mathscr{G}'(s)$ should be strictly increasing, so that $\mathscr{G}(|Du|)$ must be a symmetric strictly convex function of Du. In particular, for the area integrand $\mathscr{G}(s) = \sqrt{1+s^2} - 1$ we have $A(s) = 1/\sqrt{1+s^2}$.

In what follows, by a *classical solution* (more precisely, a *classical distribution solution*) of (1.1.2) in Ω, we mean a function $u \in C^1(\Omega)$ which satisfies (1.1.2) in the distribution sense.

In order to state the Strong Maximum Principle for the inequality (1.1.2), we shall need a further definition. With the notation $\Phi(s) = sA(s)$ when $s > 0$, and $\Phi(0) = 0$, we introduce the function

$$H(s) = s\Phi(s) - \int_0^s \Phi(s)ds, \qquad s \geq 0. \qquad (1.1.4)$$

This is easily seen to be strictly increasing, as follows from the inequality

$$s_1\Phi(s_1) - s_0\Phi(s_0) > (s_1 - s_0)\Phi(s_1) > \int_{s_0}^{s_1} \Phi(s)ds$$

when $s_1 > s_0 \geq 0$.

For the Laplace operator, the p-Laplace operator and the mean curvature operator, respectively, we have $H(s) = \frac{1}{2}s^2$, $H(s) = (p-1)s^p/p$ and

$H(s) = 1 - 1/\sqrt{1 + s^2}$. In the last example, note the anomalous behavior $\Phi(\infty) = H(\infty) = 1$, a possibility which occasionally requires extra care in the statement and treatment of results. Finally, for the variational problem (1.1.3) one has $H(s) = s\mathscr{G}'(s) - \mathscr{G}(s)$, the pre-Legendre transform of $\mathscr{G}$.

By the strong maximum principle for (1.1.2) we mean the statement that if u is a non-negative classical solution of (1.1.2) with $u(x_0) = 0$ at some point $x_0 \in \Omega$, then $u \equiv 0$ in Ω.

Theorem 1.1.1 (Strong Maximum Principle). *In order for the strong maximum principle to hold for (1.1.2) it is necessary and sufficient that either $f \equiv 0$ in $[0, d]$, $d > 0$, or that $f(s) > 0$ for $s \in (0, \delta)$ and*

$$\int_{0+} \frac{ds}{H^{-1}(F(s))} = \infty. \tag{1.1.5}$$

The choice of the base level zero for the statement of the principle is of course a matter only of convenience, as is whether we deal with minimum or maximum values at the base point x_0.

In the next result we consider the situation when the integral in (1.1.5) is convergent. Here the appropriate hypotheses are that u satisfies the converse inequality

$$\operatorname{div}\{A(|Du|)Du\} - f(u) \geq 0, \tag{1.1.6}$$

and also "*vanishes*" at ∞, rather than at some finite point $x_0 \in \Omega$. We formalize this in the following definition.

By the compact support principle for (1.1.6) we mean the statement that if u is a non-negative classical solution of (1.1.6) in an exterior domain Ω, with $u(x) \to 0$ as $|x| \to \infty$, then u has compact support in Ω.

Theorem 1.1.2 (Compact Support Principle). *Let $f(s) > 0$ for $0 < s < \delta$. Then in order for the compact support principle to hold for (1.1.6), it is necessary and sufficient that*

$$\int_{0+} \frac{ds}{H^{-1}(F(s))} < \infty. \tag{1.1.7}$$

If Theorem 1.1.2 were an exact analogue of Theorem 1.1.1, the conclusion would be that $u \equiv 0$ in Ω, but this would be incorrect since (1.1.6) admits non-negative, non-trivial compact support solutions under assumption (1.1.7), see Theorem 4.3.3.

The existence of compact support solutions for quasilinear equations was studied extensively in the 1980s, as well as other properties of the set where the solution vanishes. In chemical models, for example, when the values of a solution represent the density of a reactant, the vanishing of a solution then delineates a region, called the *dead core*, where no reactant is present (see [5], [6], [29], [81], [82], [113]). In Section 8.4 we give an extended discussion of this phenomenon.

The results described above can be extended to a wider class of differential inequalities by replacing div $\{A(|Du|)Du\}$ in (1.1.2) or (1.1.6) by the more general operator

$$\partial_{x_i}\{a_{ij}(x, u)A(|Du|)D_j u\}$$

(the obvious summation convention being used) and $f(u)$ by $-B(x, u, Du)$. Here $[a_{ij}(x, u)]$ is a continuously differentiable positive-definite symmetric matrix on $\Omega \times \mathbb{R}_0^+$, and B is continuous and satisfies

$$-\mathrm{Const.}\,\Phi(|\boldsymbol{\xi}|) - g(u) \leq B(x, u, \boldsymbol{\xi}) \leq \mathrm{Const.}\,\Phi(|\boldsymbol{\xi}|) - f(u) \qquad (1.1.8)$$

for $x \in \Omega$, $u \geq 0$ and all $\boldsymbol{\xi} \in \mathbb{R}^n$ with $|\boldsymbol{\xi}| \leq 1$, and with f and g obeying (F1) and (F2). See Theorems 5.4.1 and 5.6.1.

Some special cases of the above results are worth specific note. In particular, when $\Delta_p u - u^q \leq 0$, $p > 1$, $q > 0$, the strong maximum principle holds if and only if $q \geq p-1$, while the compact support principle holds for $\Delta_p u - u^q \geq 0$ if and only if $0 < q < p-1$. Moreover, by the main results of Section 8.4 below, there exist C^2 non-negative radially symmetric compact support solutions of $\Delta_p u - u^q = 0$ when $0 < q < p - 1$, this being an explicit case of the earlier comment after Theorem 1.1.2.

When $q = 0$ the above analysis cannot be applied. Indeed the equation $\Delta u - 2n = 0$ in any domain Ω containing the origin admits the non-trivial solution $u(x) = |x|^2$, but $u(0) = 0$. We also note that the equation $\Delta u - c = 0$, with $c \neq 0$, admits no non-negative compact support solutions for any $c \in \mathbb{R}$, as follows from the Hopf maximum principle.

An important prototype of the general situation is the equation

$$\Delta_p u - |Du|^q - f(u) = 0, \qquad p > 1, \; q > 0. \qquad (1.1.9)$$

With $\Phi(s) = s^{p-1}$ for this case, condition (1.1.8) applies with $f = g$ and requires $q \geq p - 1$. In turn, the strong maximum principle holds for (1.1.9) when $q \geq p - 1$ and either $f \equiv 0$ in $[0, d]$, $d > 0$, or f obeys (1.1.5).

On the other hand, when $q \in (0, p-1)$ the strong maximum principle can fail, even when $f \equiv 0$, e.g., the C^1 function $u(x) = C|x|^k$ satisfies

$$\Delta_p u - |Du|^q = 0, \tag{1.1.10}$$

where

$$k = \frac{p-q}{s}, \qquad \frac{1}{C} = k \left[\frac{(p-1)n - (n-1)q}{s} \right]^{1/s}, \qquad s = p - 1 - q > 0$$

(for $p = 2$, this example is due to Barles, Diaz and Diaz [8]; for general $p > 1$ it is given in [84]). It is of further interest in connection with this example that the compact support principle can fail even if (1.1.8) is satisfied, namely when $q > p - 1$! Indeed, the function $u(x) = L|x|^{-l}$ satisfies (1.1.10) in $\Omega_R = \mathbb{R}^n \setminus \overline{B}_R$, with $l = (p-q)/t > 0$, provided that

$$q > \frac{n(p-1)}{n-1}, \qquad L = \frac{1}{l} \left[\frac{(n-1)q - (p-1)n}{t} \right]^{1/t}, \qquad t = q - p + 1.$$

As we shall see in Section 2.1, for non-singular equations the Strong Maximum Principle implies the Comparison Principle , Theorem 2.1.4. On the other hand, for singular equations, even if they are smooth, the situation is more delicate. Consider for example the equation

$$\Delta_4 u + |Du|^2 = 0, \qquad n = 2. \tag{1.1.11}$$

The Strong Maximum Principle continues to hold (see Theorem 5.4.1), while on the other hand (1.1.11) admits two unequal solutions $u \equiv 0$ and $u(x) = \frac{1}{8}(R^2 - |x|^2)$ in B_R, both with the same boundary values. Thus a comparison theorem must fail. See Section 5.6 for a discussion of the more general example

$$\Delta_p u + |Du|^{q_1} - u^{q_2} \geq 0, \qquad u \geq 0, \qquad p > 1, \ q_1, q_2 > 0.$$

Chosen from among the many available applications of the maximum principle, the final chapter includes recent applications to Liouville theorems and dead core problems, and to differential inequalities on Riemannian manifolds. In Section 8.2 we also give various radial symmetry theorems for the semilinear Laplace–Poisson equation $\Delta u + f(u) = 0$ and for the quasilinear divergence structure equation

$$\operatorname{div}\{A(u, |Du|)Du\} + f(u, |Du|) = 0$$

under mild Lipschitz continuity or monotonicity conditions on the function
f. The more delicate symmetry question for over-determined boundary
value problems is treated in Section 8.3. There are of course further applications of general interest, for example Phragmèn–Lindelöf type theorems
and special Harnack theorems; the reader can be referred particularly to
[38], [76], [114] and the Notes to Chapter 7, and to recent work of Marcus
and Véron. The maximum principle can also be applied to obtain gradient
bounds for solutions of elliptic equations, using "barrier methods" or, alternatively, the application of "P-functions". For barrier methods, one can
consult [43], Chapter 14 and, for the P-function approach, the monograph
of Sperb [104].

It is beyond the scope of this book to consider fully nonlinear equations in any detail. To do this would minimally require the development and
presentation of the techniques of Krylov and Safonov to obtain Harnack
inequalities for non-divergence second order linear equations, as well as the
concept of viscosity solutions. This would altogether change our focus and
require a lengthy treatment of its own to cover the very large literature
which has grown up in this direction. The reader however can be referred
to the survey works [52], [53] and [17].

To conclude the introduction it is worth noting some further examples
of second order elliptic equations of physical and geometric interest.

1. *The equation of prescribed mean curvature:*

$$(1 + |Du|^2)\Delta u - \partial_{x_i} u\, \partial_{x_j} u\, \partial^2_{x_i x_j} u = n\mathcal{H}(x)(1 + |Du|^2)^{3/2}, \quad (1.1.12)$$

or, equally, in divergence form,

$$\mathrm{div}\left(\frac{Du}{\sqrt{1 + |Du|^2}}\right) = n\mathcal{H}(x), \qquad (1.1.13)$$

where $\mathcal{H}$ is the mean curvature of the non-parametric surface $x_{n+1} = u(x)$
in the $(n+1)$-dimensional (x, x_{n+1})-space. This equation arises naturally by
considering the isoperimetric problem of least surface area bounding a given
volume; it had already been derived by Lagrange in 1760. Of additional
interest is the case when $\mathcal{H}$ is specified as a function of x, u and Du. Some
special examples of this type occur below.

2. *The surface of a fluid under the combined action of gravity and surface
tension (capillary surface)*

$$(1 + |Du|^2)\Delta u - \partial_{x_i} u\, \partial_{x_j} u\, \partial^2_{x_i x_j} u = \kappa\, u(1 + |Du|^2)^{3/2},$$

where κ is an appropriate physical constant. In the physically central case of two dimensions this equation arises from balancing forces of tension (proportional to the mean curvature of the capillarity surface) with the weight of the fluid supported. The constant κ is positive or negative depending on whether the surface in question is an upper or lower boundary of the fluid.

3. *Central projection.*Let S^n be the sphere of $\mathbb{R}^{n+1}$, which can be mapped conformally onto the Euclidean tangent space $\mathbb{R}^n$ at the South Pole by means of stereographic projection from the North Pole. In this projection the volume element is $dm = dx/(1 + |x|^2)^n$ and the gradient ∇ on S^n is expressed as $(1 + |x|^2)D$, where D stands for the Euclidean gradient in $\mathbb{R}^n$ and x denotes a Euclidean coordinate centered at the South Pole.

As a particular example, the p-Dirichlet norm on S^n, $p > 1$, is then minimized by functions u on S^n which satisfy

$$\mathrm{div}_{S^n}(|\nabla u|^{p-2}\nabla u) = 0.$$

Reverting to the stereographic variable x this has the form

$$\rho^{-n}\mathrm{div}(\rho^{n-p}|Du|^{p-2}Du) = 0, \qquad \rho(x) = 1/(1 + |x|^2),$$

this being a particular example where the vector $\boldsymbol{A}$ depends on both x and Du. Of course, general variational integrals on S^n can be treated in the same way.

4. *Subsonic gas dynamics.*The velocity potential φ satisfies

$$\mathrm{div}(\varrho D\varphi) = 0,$$

where the velocity $D\varphi$ and the density ϱ are related through Bernoulli's law. For the important case of an ideal gas the relation is

$$\frac{1}{2}|D\varphi|^2 + \frac{c^2}{\gamma - 1} = \mathrm{Const.}, \qquad c = \text{sound spead} \sim \varrho^{(\gamma-1)/2},$$

where $\gamma > 1$ is the ratio of the specific heats of the gas.

5. *The general equation of radiative cooling*

$$\mathrm{div}(\kappa|Du|^{p-2}Du) = \sigma u^4, \qquad p > 1,$$

where κ is the coefficient of heat conduction, depending on x and possibly also on u, while σ is the radiation, assumed to be constant. Replacing the right-hand side by various functions $f = f(x, u)$ yields further examples of physical interest.

6. *The Euler–Lagrange equation.* For the variational problem

$$\delta \int_\Omega \mathscr{G}(x, u, Du)\, dx = 0,$$

with $\mathscr{G} = \mathscr{G}(x, z, \boldsymbol{\xi})$ being of class C^1, this takes the form

$$\operatorname{div} \partial_{\boldsymbol{\xi}} \mathscr{G}(x, u, Du) = \partial_z \mathscr{G}(x, u, Du).$$

Ellipticity is equivalent to strong convexity of $\mathscr{G}$ with respect to $\boldsymbol{\xi}$, namely the figuratrix surface $x_{n+1} = \mathscr{G}(x, z, \boldsymbol{\xi})$ should have positive Gaussian curvature for fixed (x, z).

If $\mathscr{G}$ is jointly convex in z and $\boldsymbol{\xi}$ and satisfies mild regularity conditions, then the solution of the Euler–Lagrange equation provides a minimizing function for the variational problem.

The case where $\mathscr{G}$ depends only on z and $|\boldsymbol{\xi}|$ is particularly to be noted since the corresponding problem is invariant under rotations of the underlying space.

7. *The 2-dimensional Monge–Ampére equation*

$$e(\partial_{x^2}^2 u\, \partial_{y^2}^2 u) + a\, \partial_{x^2}^2 u + 2b\, \partial_{xy}^2 u + c\, \partial_{y^2}^2 u = d$$

is elliptic if and only if $ac - b^2 + ed > 0$. Here a, b, c, d, e depend on (x, y), or more generally on $(x, y, u, \boldsymbol{\xi})$, $\boldsymbol{\xi} = (\xi_1, \xi_2)$.

8. *Calabi's equation*

$$\det D^2 u = f(x).$$

Ellipticity demands that the surface $x_{n+1} = u(x)$ be convex.

1.2   Notation

Throughout, we shall let $x = (x_1, \ldots, x_n)$ denote points of $\mathbb{R}^n$, $n \geq 1$, and will denote the solution variable by $u = u(x)$. We put as before $\partial u / \partial x_i = \partial_{x_i} u$, $\partial^2 u / \partial x_i \partial x_j = \partial_{x_i x_j}^2 u$ when the solutions are assumed to be classical, that is of class C^2 in any domain of interest. We also write $Du = \operatorname{grad} u = (\partial_{x_1} u, \ldots, \partial_{x_n} u)$ for the gradient vector of u, and $D^2 u = [\partial_{x_i x_j}^2 u]$ for the Hessian matrix of u. It is understood that repeated subscripts i, j, k etc. are summed over the appropriate range indicated by the context.

A domain Ω in $\mathbb{R}^n$ is always understood to be a connected open set in $\mathbb{R}^n$; thus any open and relatively closed non-empty subset coincides with Ω itself. We denote the boundary of Ω by $\partial\Omega$, and the closure of Ω by $\overline{\Omega}$. By $\Omega' \subset\subset \Omega$ we mean that Ω' is a subdomain with compact closure in Ω. The notation $\langle \cdot, \cdot \rangle$ is always reserved for the inner product in the (vector) space $\mathbb{R}^n$.

We assume the reader to have a standard background in real analysis including Sobolev spaces, but without need for linear operator theory. A useful assortment of classical results and techniques can be found in [43], Sections 7.1–7.7.

Chapter 2

Tangency and Comparison Theorems for Elliptic Inequalities

2.1 The contributions of Eberhard Hopf

We begin with the classical maximum principle due to E. Hopf [46], together with an extended commentary and discussion of Hopf's original paper by J. Serrin [97].

The maximum principle for harmonic and subharmonic functions was known to Gauss on the basis of the mean value theorem (1839); an extension to elliptic inequalities however remained open until the twentieth century. Bernstein (1904), Picard (1905), Lichtenstein (1912, 1924) then obtained various results by difficult means, as well as use of regularity conditions for the coefficients of the highest order terms. Moreover, a few months before Hopf's paper, there appeared an article of Picone [71] containing ideas similar to Hopf's, but with weaker conclusions. It was Hopf's genius to see that a *"gänzlich elementares Begründen"* could be given. The comparison technique he invented for this purpose is essentially so transparent that it has generated important applications in many further directions.

Here is Hopf's theorem in its main form:

Hopf's Maximum Principle. *Let* $u = u(x)$, $x = (x_1, \ldots, x_n)$, *be a* C^2 *function which satisfies the differential inequality*

$$Lu \equiv \sum_{i,j} a_{ij}(x)\partial^2_{x_i x_j} u + \sum_i b_i(x)\partial_{x_i} u \geq 0$$

in a domain Ω. *Suppose the (symmetric) matrix* $[a_{ij}] = [a_{ij}(x)]$ *is locally uniformly positive definite in* Ω *(that is, for any given compact subset* Ω' *of* Ω, *the quadratic form*

$$\sum_{i,j} a_{ij}(x)\eta_i\eta_j$$

is positive and uniformly bounded from 0 *for all* x *in* Ω' *and all vectors* $\boldsymbol{\eta}$ *in* $\mathbb{R}^n$ *with* $|\boldsymbol{\eta}| = 1$), *and the coefficients* a_{ij}, $b_i = b_i(x)$ *are locally bounded in* Ω.

If u *takes a maximum value* M *in* Ω, *then* $u \equiv M$ *in* Ω.

Hopf's proof (Section I of [46]), now a classic of the subject, is reproduced in the monographs [76], [43], [38] and in many other texts as well, particularly the second volume of [22]. We give a proof in the Appendix of this chapter, Section 2.8.

The hypothesis that u *is twice differentiable is essential for the theorem, though not always strictly noted in presentations of the result.*

In Section II of [46] Hopf notices two important corollaries (Sätze 2, 3) dealing with the differential inequality $Lu + c(x)u \geq 0$. *First, for the case* $c = c(x) \leq 0$ *and a positive maximum, and second, when there is an extremum* $M = 0$ *irrespective of the sign of* c. The latter possibility is mentioned only in passing in [43], and not at all in Courant and Hilbert [22]. The formal statement of these corollaries is as follows.

Theorem 2.1.1. *Let* u *be a* C^2 *function satisfying the differential inequality*

$$Lu + c(x)u \geq 0 \quad (\leq 0) \tag{2.1.1}$$

in a domain Ω, *where the coefficients of* L *satisfy the previous conditions, and* $c = c(x) \leq 0$ *in* Ω. *If* u *takes a positive maximum (negative minimum) value* M *in* Ω, *then* $u \equiv M$.

The result is easy to prove. That is, near a positive maximum M of u we would have

$$Lu \geq -c(x)u \geq 0.$$

Hopf's main theorem then yields $u \equiv M$ near the maximum point; in turn $u \equiv M$ in all Ω (the set $\{x \in \Omega : u = M\}$ is non-empty and both open and closed in the connected set Ω).

Hopf's second result is

Theorem 2.1.2. *Let the hypotheses of Theorem 2.1.1 hold, except that one now assumes alternatively that the function c is locally bounded below in Ω. If u takes on a vanishing maximum (minimum) value $M = 0$ in Ω, then $u \equiv 0$.*

Proof. (Hopf.) Let $u \leq 0$ in Ω and define $v(x) = e^{-\alpha x_1} u(x)$, $x \in \Omega$, $\alpha > 0$. Clearly $v \in C^2(\Omega)$, is non-positive and satisfies the differential inequality

$$Lv + \sum_i \tilde{b}_i(x)\, \partial_{x_i} v \geq -\tilde{c}(x)\, e^{-\alpha x_1} u, \qquad \tilde{c} = c + \alpha^2 a_{11} + \alpha b_1,$$

where $\tilde{b}_i = 2\alpha a_{i1}$. In any domain Ω' with compact closure in Ω we have

$$c(x) \geq -\text{const.}, \quad |b_1(x)| \leq \text{const.}, \quad a_{11}(x) \geq \text{const.} > 0.$$

Therefore we can choose α sufficiently large so that $\tilde{c}(x)$ is positive in Ω'. In turn $Lv + \sum_i \tilde{b}_i \partial_{x_i} v \geq 0$ in Ω'. Let $y \in \Omega$ be such that $u(y) = 0$ and take Ω' containing y. Then $v(y) = 0$ and by Hopf's main theorem we get $u \equiv v \equiv 0$ in Ω', from which it follows at once that $u \equiv 0$ in the entire Ω. The case $u \geq 0$ in Ω is treated in the same way. $\qquad \square$

It may be remarked that earlier statements of Theorems 2.1.1 and 2.1.2 have usually imposed stronger boundedness conditions on the function $c(x)$ than those required here. Observe also that Theorem 2.1.2 can be slightly generalized as follows:

Theorem 2.1.2′. *Let $u \in C^2(\Omega)$ satisfy*

$$\sum_{i,j} a_{ij}\partial^2_{x_i x_j} u \leq b(x)(u + |Du|),$$

with a_{ij}, b locally bounded in Ω, and a_{ij} locally uniformly positive definite. If $u \geq 0$ in Ω and u is zero at some point x_0 in Ω, then $u \equiv 0$ in Ω.

We omit the proof (see Problem 2.3).

As is customary, the term *strong maximum principle* will be used here to denote the main results of Hopf stated above, as well as related results, e.g., Theorem 2.1.1. On the other hand, the term *maximum principle* (in contrast to strong maximum principle) is reserved to denote results in which a bound for a solution u of an elliptic equation, or inequality, is given in terms of an a priori bound for u on the boundary of its domain of definition. This terminology follows, e.g., Gilbarg and Trudinger [43] and Fraenkel [38].

Continuing with the discussion of Hopf's work, in Section II of [46] Hopf observes that one can allow the coefficients $a_{ij}(x)$, $b_i(x)$, $c(x)$ to depend on the solution u itself, provided that when they are evaluated along the solution the resulting functions $\tilde{a}_{ij}(x)$, $\tilde{b}_i(x)$, $\tilde{c}(x)$ satisfy the conditions of the main theorems. This allows him to deal explicitly with nonlinear as well as linear equations.

The real depth of Hopf's nonlinear analysis shows up only in Section III, where he considered the *fully nonlinear equation of second order*

$$\mathscr{F}(x, u, Du, D^2u) = 0, \tag{2.1.2}$$

the structure of the equation being determined by the function $\mathscr{F}(x, z, \boldsymbol{\xi}, \boldsymbol{s})$, where z, $\boldsymbol{\xi}$ and $\boldsymbol{s}$ are respectively placeholders for u, Du and D^2u. Here (2.1.2) is said to be *elliptic* if the matrix $D_{\boldsymbol{s}}\mathscr{F}$ is positive definite for all relevant values of its variables.

Hopf's presentation is, however, seriously obscured by the restriction to exact equations, rather than corresponding differential inequalities as in the preceding results, as well as to the case where one of the solutions in question is assumed to vanish identically ("*engere Voraussetzungen*" according to Hopf). Accordingly we shall restate the results in slightly greater generality and in more usual notation. Hopf's first result is a beautiful *tangency principle*, essentially Satz 3' of [46].

Theorem 2.1.3 (Tangency Principle). *Let* u, v *be* $C^2(\Omega)$ *solutions of the nonlinear differential inequality*

$$\mathscr{F}(x, u, Du, D^2u) \geq \mathscr{F}(x, v, Dv, D^2v),$$

where the function $\mathscr{F} = \mathscr{F}(x, z, \boldsymbol{\xi}, \boldsymbol{s})$ *is continuously differentiable in the variables* z, $\boldsymbol{\xi}$, $\boldsymbol{s}$, *that is, the derivatives* $\partial_z\mathscr{F}$, $\partial_{\boldsymbol{\xi}}\mathscr{F}$, $\partial_{\boldsymbol{s}}\mathscr{F}$ *exist and are continuous functions of* $(x, z, \boldsymbol{\xi}, \boldsymbol{s}) \in \Omega \times \mathbb{R} \times \mathbb{R}^n \times \mathbb{R}^{n \times n}$. *Suppose also that the matrix* $Q = [Q_{ij}]$ *given by*

$$Q_{ij} \equiv \partial_{\boldsymbol{s}}\mathscr{F}(x, u, Du, \theta D^2u + (1 - \theta)D^2v),$$

is positive definite in $x \in \Omega$ *and all* $\theta \in [0, 1]$.

If $u \leq v$ in Ω and $u = v$ at some point x_0 in Ω, then $u \equiv v$ in Ω. The terms u, Du in Q can be replaced by v, Dv.

Proof. Essentially following Hopf's proof of Satz $3'$ of [46], we write

$$0 \geq \mathscr{F}(x, v, Dv, D^2v) - \mathscr{F}(x, u, Du, D^2u)$$
$$= \mathscr{F}(x, u, Du, D^2v) - \mathscr{F}(x, u, Du, D^2u) + \mathscr{F}(x, u, Dv, D^2v)$$
$$\quad - \mathscr{F}(x, u, Du, D^2v) + \mathscr{F}(x, v, Dv, D^2v) - \mathscr{F}(x, u, Dv, D^2v)$$
$$= \sum a_{ij} \partial^2_{x_i x_j} (v - u) + \sum b_i \partial_{x_i} (v - u) + c(v - u)$$
$$\equiv L(v - u) + c(v - u),$$

where, for some values θ, θ_1, $\theta_2 \in [0, 1]$, depending on x, we have by the mean value theorem

$$a_{ij} = \partial_s \mathscr{F}(x, u, Du, \theta D^2 v + (1 - \theta)D^2 u) = Q_{ij}|_{\theta=\theta(x)},$$
$$b_i = \partial_{\xi_i} \mathscr{F}(x, u, \theta_1 Dv + (1 - \theta_1)Du, D^2 v)|_{\theta_1=\theta_1(x)},$$
$$c = \partial_z \mathscr{F}(x, \theta_2 v + (1 - \theta_2)u, Dv, D^2 v)|_{\theta_2=\theta_2(x)}.$$

Since Q_{ij} is continuous for $x \in \Omega$ and $\theta \in [0, 1]$, the principal condition on Q_{ij} shows that in fact it is *uniformly* positive definite for $x \in \Omega'$ and $\theta \in [0, 1]$, when Ω' is a compact subset of Ω. Consequently the coefficient matrix $[a_{ij}]$ is locally uniformly positive definite on Ω. By the same argument it is clear that also a_{ij}, b_i, c are locally bounded in Ω. Since by assumption $v - u \geq 0$ and $(v - u)(x_0) = 0$, it now follows from Theorem 2.1.2 that $v \equiv u$ in Ω.

To obtain the final conclusion of the theorem, one proceeds in the same way, though starting from the alternative decomposition

$$0 \geq \mathscr{F}(x, v, Dv, D^2v) - \mathscr{F}(x, u, Du, D^2u)$$
$$= \mathscr{F}(x, v, Dv, D^2v) - \mathscr{F}(x, v, Dv, D^2u) + \mathscr{F}(x, v, Dv, D^2u)$$
$$\quad - \mathscr{F}(x, v, Du, D^2u) + \mathscr{F}(x, v, Du, D^2u) - \mathscr{F}(x, u, Du, D^2u),$$

but otherwise leaving the proof unchanged. $\qquad\square$

Hopf's Theorems 2.1.1 and 2.1.2 are in fact tangency principles in which the second solution v is constant ($= M$).

The next result (essentially Satz $2'$ of [46] in a more general context and formulation) is stated here as a *comparison result*, rather than a maximum principle, this being the underlying content of Hopf's theorem. By

$u \leq v$ on $\partial\Omega$ we mean explicitly that for every $\delta > 0$ there is a neighborhood of $\partial\Omega$ in which $u \leq v + \delta$.

Theorem 2.1.4 (Comparison Principle). *Let u, v be $C^2(\Omega)$ solutions of the nonlinear differential inequality given in Theorem 2.1.3. Suppose that the matrix $Q = [Q_{ij}]$ is positive definite in Ω and that for every fixed $x \in \Omega$ the function*

$$\mathscr{F}(x, \cdot, Dv(x), D^2v(x)) : \mathbb{R} \to \mathbb{R} \tag{2.1.3}$$

is non-increasing on the semi-line $[v(x), \infty)$ – but not necessarily differentiable. If $u \leq v$ on $\partial\Omega$, then $u \leq v$ in Ω.[1]

The terms u, Du in Q can be replaced by v, Dv if at the same time the terms Dv, D^2v in (2.1.3) are replaced by Du, D^2u and the semi-line $[v(x), \infty)$ is replaced by $(-\infty, u(x)]$.

Proof. Suppose for contradiction that the conclusion $v - u \geq 0$ in Ω fails.

Then there will be a subdomain Ω' of Ω in which $v - u < 0$ but is not identically constant, and in which also $v - u$ takes on a negative minimum M at a point y. As in the proof of Theorem 2.1.3, one obtains with the help of (2.1.3) that $L(v - u) \leq 0$ in Ω', where L has the obvious meaning. Hence by Hopf's main theorem we get $v - u \equiv M$ in Ω', a contradiction.

The final conclusion is obtained from the alternative decomposition in the proof of Theorem 2.1.3. □

Using other decompositions, one can obtain various related results, see, e.g., Theorem 31 of Chapter 2 of [76].

A direct consequence of Theorem 2.1.4 is a uniqueness theorem for the Dirichlet problem for the nonlinear equation $\mathscr{F}(x, u, Du, D^2u) = 0$, a fact mentioned by Hopf in the final paragraph of [46], though not explicitly formulated by him. Since the result is important, and a precise formulation is in fact not immediate from Hopf's analysis, it is worth stating the definite result here.

Theorem 2.1.5. *Let u and v be $C^2(\Omega)$ solutions of the nonlinear equation*

$$\mathscr{F}(x, u, Du, D^2u) = 0 \tag{2.1.4}$$

in a domain Ω, with $u = v$ on $\partial\Omega$. Suppose Q is positive definite in Ω for all $\theta \in [0, 1]$, and that $\mathscr{F}(x, \cdot, Dv(x), D^2v(x))$ is non-increasing on the entire line $\mathbb{R}$; see (2.1.3). Then $u \equiv v$.

---

[1] In fact by Theorem 2.1.3, if $\partial_z \mathscr{F}$ is also continuously differentiable, then either $u \equiv v$ in Ω or $u < v$ in Ω.

This is an immediate corollary of Theorem 2.1.4, the main result being used to establish that $u \leq v$, and the final part used to get $v \leq u$. Here it is crucial that (2.1.3) holds on the *entire line* $\mathbb{R}$.

It is surprising that the matrix Q in the hypothesis of Theorem 2.1.5 is, insofar as its second and third arguments are concerned, to be evaluated solely on the functions u and Du, *without any symmetric reference to v and Dv*.

The maximum principle, simple enough in essence, nevertheless lends itself to a quite remarkable number of uses when combined appropriately with other notions. We discuss several here, reserving more subtle applications until the final chapter of the book.

A general quasilinear equation of second order, for example, has the form

$$\boldsymbol{a}(x, u, Du)D^2 u + B(x, u, Du) = 0, \quad x \in \Omega, \tag{2.1.5}$$

where $\boldsymbol{a} = \boldsymbol{a}(x, z, \boldsymbol{\xi})$ and $B = B(x, z, \boldsymbol{\xi})$ are respectively a given $n \times n$ matrix $[a_{ij}]$ and a given scalar function of the variables $(x, z, \boldsymbol{\xi}) \in \Omega \times \mathbb{R} \times \mathbb{R}^n$. The notation $\boldsymbol{a} \, D^2 u$ denotes the natural contraction $\sum_{i,j} a_{ij}\partial^2_{x_i x_j} u$.

A classical solution $u \in C^2(\Omega)$ of (2.1.5) is called *elliptic* if the matrix $\boldsymbol{a}(x, u, Du)$ is positive definite when evaluated at $u = u(x)$, $x \in \Omega$. The equation itself is called elliptic in Ω, or simply elliptic, if $\boldsymbol{a}(x, z, \boldsymbol{\xi})$ is positive definite for all $(x, z, \boldsymbol{\xi}) \in \Omega \times \mathbb{R} \times \mathbb{R}^n$.

In view of Theorem 2.1.5, a sufficient condition for uniqueness of the corresponding Dirichlet problem for (2.1.5), with $u \in C^2(\Omega) \cap C(\overline{\Omega})$ and u given on $\partial\Omega$, is that *the matrix $\boldsymbol{a}$ is independent of z, the scalar function $B(x, z, \boldsymbol{\xi})$ is non-increasing in z for arbitrary arguments x, $\boldsymbol{\xi}$, and there exists at least one (!) elliptic solution u*. This conclusion is essentially due to Hopf, though not explicitly mentioned or stated by him; it seems to have appeared first in [43], first edition, Chapter 8.

This result applies at once to the quasilinear operator

$$(1 + |Du|^2)\Delta u - \sum_{i,j} \partial_{x_i} u \, \partial_{x_j} u \, \partial^2_{x_i x_j} u$$

(mean curvature) for which the corresponding matrix

$$Q_{ij} = a_{ij} = (1 + |Du|^2)\delta_{ij} - \partial_{x_i} u \partial_{x_j} u$$

is positive definite for all values of its arguments (that is, the mean curvature operator is elliptic). Here of course there is no need to use the full

strength of Theorem 2.1.5. On the other hand, if we consider the Dirichlet
problem

$$(1 + |Du|^2)\Delta u - 2 \sum_{i,j} \partial_{x_i} u \, \partial_{x_j} u \, \partial^2_{x_i x_j} u = 0$$

in Ω, with $u = 0$ on $\partial\Omega$, then the matrix is *not* positive definite for arbitrary
arguments $D^2 u$. Nevertheless $Q = I\!I$ for the function $u \equiv 0$, whence it
follows that 0 is the *unique* solutionof the Dirichlet problem.

A second and more subtle example is the elementary Monge–Ampère
equation in $\mathbb{R}^2$,

$$\partial^2_{x^2} u \, \partial^2_{y^2} u - \left(\partial^2_{xy} u\right)^2 = g(x, y).$$

Here one checks that

$$Q_{ij}\xi_i\xi_j = \partial^2_{y^2} u \, \xi_1^2 - 2\, \partial^2_{xy} u \, \xi_1\xi_2 + \partial^2_{x^2} u \, \xi_2^2.$$

The determinant of Q, $\det Q$, is then equal to $\det \mathbb{H}u = \partial^2_{x^2} u \, \partial^2_{y^2} u -$
$\left(\partial^2_{xy} u\right)^2$, which is precisely $g = g(x, y)$ when evaluated at a solution u.

Suppose in particular that $g > 0$. It is easy to see then, that *any solution u is either everywhere strictly convex or everywhere strictly concave.*

From this, one can check without difficulty that *if u and v are two
convex solutions, then Q is positive definite for the arguments $\partial^2_{x_i x_j}(\theta u +
(1 - \theta)v)$.*

Hence the Dirichlet problem for the elementary Monge–Ampère equation above has at most one convex solution. On the other hand, if u and
v are concave solutions, then $-u$ and $-v$ are convex solutions and so,
similarly, the Dirichlet problem can have at most one concave solution;
altogether then *the problem can have at most two solutions.* This result is
a special case of a theorem of Rellich [88]; see [22, page 324].

Other related maximum and comparison principles are discussed in
the Notes to Chapter 2 of [76], to which the reader is strongly referred;
see also the references cited on page 314 of [114]. Several recent maximum
principles for singular fully nonlinear equations are given in [7], [14], based
on the "*viscosity*" method.

Hopf's proof technique, as noted above, leads to other results of fun-
damental interest, particularly the celebrated Boundary Point Lemma and
a Harnack principle for nonlinear elliptic equations in two variables, see
Theorem 2.8.3, [83, Section 5.5] and [43, Chapter 3].

2.2 Tangency and comparison principles for quasilinear inequalities

We consider the pair of differential inequalities

$$a_{ij}(x, u, Du)\partial^2_{x_i x_j} u + B(x, u, Du) \geq 0, \qquad (2.2.1)$$

$$a_{ij}(x, v, Dv)\partial^2_{x_i x_j} v + B(x, v, Dv) \leq 0, \qquad (2.2.2)$$

where the standard summation convention is assumed to be in effect. Let P be an open subset of $\mathbb{R}^n$ and let the matrix of coefficients

$$[a_{ij}] = [a_{ij}(x, z, \boldsymbol{\xi})] : K \to \mathbb{R}^{n^2}, \qquad K = \Omega \times \mathbb{R} \times \boldsymbol{P},$$

be continuous, and also continuously differentiable with respect to z and $\boldsymbol{\xi}$, in the set K. Moreover, let $B = B(x, z, \boldsymbol{\xi}) : K \to \mathbb{R}$ be locally Lipschitz continuous with respect to $\boldsymbol{\xi}$ in K.

The set $\boldsymbol{P}$ is called the *regular set*, while $\boldsymbol{Q} = \mathbb{R}^n \setminus \boldsymbol{P}$ is the *singular set* for (2.2.1) and (2.2.2). It is not necessary that the inequalities (2.2.1) and (2.2.2) even have meaning for points x in Ω for which $Du(x)$ or $Dv(x)$ are in the singular set. These conditions apply in particular to the p-Laplace operator Δ_p, where

$$[a_{ij}] = [a_{ij}(\boldsymbol{\xi})] = |\boldsymbol{\xi}|^{p-2}\left[\boldsymbol{I} + (p-2)\frac{\boldsymbol{\xi} \otimes \boldsymbol{\xi}}{|\boldsymbol{\xi}|^2}\right], \qquad \boldsymbol{\xi} \neq \boldsymbol{0};$$

this is singular when $p \neq 2$, with the singular set $\boldsymbol{Q} = \{\boldsymbol{0}\}$. (The matrix $[a_{ij}]$ is even undefined at $\boldsymbol{\xi} = \boldsymbol{0}$ when $p < 2$.)

The inequalities (2.2.1), (2.2.2) are called *elliptic* if $a(x, z, \boldsymbol{\xi})$ is positive definite for $(x, z, \boldsymbol{\xi}) \in \Omega \times \mathbb{R} \times \boldsymbol{P}$. Similarly, a solution v of (2.2.2) is called *elliptic* if the matrix $a(x, v, Dv)$ is positive definite when evaluated at $v = v(x)$, $x \in \Omega$. The corresponding terminology applies of course to solutions of (2.2.1).

Theorem 2.2.1 (Tangency Principle). *Let v be an elliptic solution of (2.2.2) in Ω, with $Dv(x) \in \boldsymbol{P}$ for all $x \in \Omega$, and u be a solution of (2.2.1) in Ω, of class C^2 in the open set $U = \{x \in \Omega : Du(x) \in \boldsymbol{P}\}$, where $\boldsymbol{P}$ is the regular set for the inequalities (2.2.1) and (2.1.2).*

Assume moreover that $B(x, z, \boldsymbol{\xi})$ is locally lower Lipschitz continuous with respect to the variable z in K.[2]

If $u \leq v$ in Ω and $u = v$ at some point $x_0 \in \Omega$, then $u \equiv v$ in Ω.

---

[2] That is, for every compact subset of K there is a number $b_2 > 0$ such that if $\bar{z} > z$,

The conclusion can be informally restated as saying that if u and v are one-sidedly tangent at a point in Ω, then they coincide. It is interesting to note that *no condition of ellipticity* is required of (2.2.1) itself. The same remark applies also to the next two theorems.

When the regular set is all of $\mathbb{R}^n$ (that is, $\boldsymbol{Q} = \emptyset$) Theorem 2.2.1, as well as later theorems, has a simpler formulation.

Theorem 2.2.2 (Tangency Principle). *Let $\boldsymbol{P} = \mathbb{R}^n$. Suppose that u and v are respectively solutions of (2.2.1) and (2.2.2) in Ω of class $C^2(\Omega)$, with v being elliptic in Ω.*

Assume also that $B(x, z, \boldsymbol{\xi})$ is locally lower Lipschitz continuous with respect to z in K. If $u \leq v$ in Ω and $u = v$ at some point in Ω, then $u \equiv v$ in Ω.

Proof. It is enough to prove Theorem 2.2.1. Let $E = \{x \in \Omega : u(x) = v(x)\}$. By assumption $E \neq \emptyset$, while of course E is closed. Fix $y \in E$. Since $w = u - v \leq 0$ in Ω and $w(y) = 0$, we have $Dw(y) = \boldsymbol{0}$.

Since $Dv(y) \in \boldsymbol{P}$ and $Du(y) = Dv(y)$, there is a suitably small $\sigma > 0$ such that $Du(x)$, $Dv(x) \in \boldsymbol{P}$ for all $x \in B_\sigma$, where $B_\sigma = B_\sigma(y)$ is the closed ball with center y and radius σ in Ω. Obviously $B_\sigma \subset U$. As in the proof of Theorem 2.1.3, but now with $\mathscr{F}(x, u, Du, D^2u) = a_{ij}(x, u, Du)\partial^2_{x_i x_j} u + B(x, u, Du)$, we obtain the inequality

$$a_{ij}(x, v, Dv)\partial^2_{x_i x_j} w + b_i(x)\partial_{x_i} w + c(x)w \geq -b\big(w + |Dw|\big) \quad \text{in } B_\sigma, \quad (2.2.3)$$

where b is a non-negative constant depending on the given conditions of Lipschitz continuity of B in z and $\boldsymbol{\xi}$, and on B_σ, while

$$b_k = \big(\partial_{\xi_k} a_{ij}(x, v, \theta_1 Du + (1 - \theta_1)Dv)\big)\partial^2_{x_i x_j} u,$$

$$c = \big(\partial_z a_{ij}(x, \theta_2 u + (1 - \theta_2)v, Du)\big)\partial^2_{x_i x_j} u$$

for some values θ_1, $\theta_2 \in [0, 1]$. Clearly a_{ij}, b_i and c are bounded in B_σ, and equally by continuity the coefficient matrix $[a_{ij}(x, v, Dv)]$ is uniformly positive definite in B_σ. Because w has a zero maximum in B_σ, it now follows from Theorem 2.1.2′ applied to the nonlinear inequality (2.2.3) that $w \equiv 0$

---

then

$$B(x, \bar{z}, \boldsymbol{\xi}) - B(x, z, \boldsymbol{\xi}) \geq -b_2(\bar{z} - z)$$

in the subset.

In the formulation of Theorem 2.2.1 the inequalities (2.2.1) and (2.2.2) could be taken in the form $Lu - Lv \geq 0$. The present formulation is equivalent and perhaps easier to visualize.

in B_σ, that is $B_\sigma \subset E$. Hence E is also an open set. By the connectedness of Ω it follows that $E = \Omega$, as required. □

Theorem 2.2.3 (Comparison Principle). *As in Theorem 2.2.1, let u and v be of class $C^2(\Omega)$ with $Dv(x) \in \boldsymbol{P}$ for all $x \in \Omega$. Suppose that u is a solution of (2.2.1) in the open set $U = \{x \in \Omega : Du(x) \in \boldsymbol{P}\}$, while v is an elliptic solution of (2.2.2) in Ω.*

Assume that $[a_{ij}]$ is independent of z and that B is non-increasing with respect to z in K. If $u \leq v$ on $\partial\Omega$, then $u \leq v$ in Ω.

Remark. The reader should note the rather different hypotheses in Theorems 2.2.1 and 2.2.3. It can be shown by example that the specific monotonicity stated for B in these results cannot be reversed.

In view of conclusion $u \leq v$ of Theorem 2.2.3, solutions of the inequalities (2.2.1) and (2.2.2) are frequently called, respectively, *subsolutions* and *supersolutions* of the equation

$$a_{ij}(x, u, Du)\partial^2_{x_i x_j} u + B(x, u, Du) = 0.$$

Proof of Theorem 2.2.3. The proof is by contradiction, essentially the same as for Theorem 2.1.4. Let Ω' be a subdomain of Ω in which $w = u - v > 0$ but is not identically constant, and in which also w takes on a positive maximum M at a point y. Obviously $Dw = \boldsymbol{0}$ at y.

Hence, as in the proof of Theorem 2.2.1, there exists a closed ball $B_\sigma \subset \Omega'$ centered at y such that $Du(x)$, $Dv(x) \in \boldsymbol{P}$ for all $x \in B_\sigma$. Clearly $B_\sigma \subset U$. Moreover, as in the proof of Theorem 2.2.1, but using the fact that $[a_{ij}]$ is independent of z and also the monotonicity of B in z, we get (see (2.2.3))

$$a_{ij}(x, Dv)\partial^2_{x_i x_j} w + b_k \partial_{x_k} w \geq -b|Dw| \quad \text{in } B_\sigma.$$

Since the inequality is invariant up to constants, it now follows from Theorem 2.1.2′ that $w \equiv M > 0$ in B_σ. The subset of Ω' where $w \equiv M$ is thus both open and relatively closed. Hence $w \equiv M$ in Ω' and this fact contradicts the definition of Ω'. □

As in the case of Theorem 2.2.1, when the regular set is all of $\mathbb{R}^n$ the proof can be considerably simplified.

Norman Meyers [59] has shown that the comparison Theorem 2.2.3 fails if the coefficient matrix $[a_{ij}]$ depends on the z variable. At the same time, by considering the function v in Theorem 2.2.3 as a "*comparison function*", the conclusion can be interpreted as a maximum principle. We take up this idea in the next section.

The next result applies to semilinear rather than quasilinear inequalities, for example

$$\Delta u + f(x, u) \geq 0, \qquad \Delta v + f(x, v) \leq 0.$$

Theorem 2.2.4 (Comparison Principle). *Let L be the linear differential operator given in Hopf's main theorem (Section 2.1), and let u, $v \in C^2(\Omega)$ be solutions of the differential inequalities*

$$Lu + f(x, u) \geq 0, \qquad Lv + f(x, v) \leq 0$$

in Ω with $v > 0$ in $\overline{\Omega}$.

Suppose that $z \mapsto f(x, z)/z$, $z > 0$, is a non-increasing function for each fixed $x \in \Omega$. Then if $u \leq v$ on $\partial\Omega$ we have $u \leq v$ in Ω.

The condition on f here is more general than simple monotonicity, as one sees from the example $f(z) = z^q$, which is non-increasing when $q \leq 0$, while $f(z)/z = z^{q-1}$ is non-increasing when $q \leq 1$.

Proof. Put $w = w(x) = u(x)/v(x)$ in Ω, so

$$Lw + \frac{2}{v} a_{ij} \partial_{x_j} v \, \partial_{x_i} w = \frac{1}{v} Lu - \frac{u}{v^2} Lv \geq \left\{ \frac{f(x, v)}{v} - \frac{f(x, u)}{u} \right\} w. \quad (2.2.4)$$

Since $u \leq v$ and $v > 0$ on $\partial\Omega$ it follows that also $w \leq 1$ on $\partial\Omega$. If the conclusion $w \leq 1$ fails at some point in Ω, there would be a point x_0 in Ω where w takes a maximum value $M > 1$. In the neighborhood of x_0 the right side of (2.2.4) would then be non-negative according to hypothesis, so by Hopf's main theorem, with b_i replaced by $b_i + (2/v)a_{ij}\partial_{x_j} v$, we would have $w \equiv M > 1$ in this neighborhood, and then $w \equiv M$ in Ω, which is impossible. $\qquad\square$

An alternate proof of Theorem 2.2.4 can be given based on the substitutions $w = \log u$, $w' = \log v$.

Example. As a consequence of this theorem, Protter and Weinberger [76] have observed that when $L = \Delta$ and $f(x, u) = 2u$, there can be no positive solutions of $\Delta v + 2v \leq 0$ in the 2-dimensional square $\Omega = \{|x| \leq \pi/2, |y| \leq \pi/2\}$. Indeed, if this were the case, then any solution of $\Delta u + 2u = 0$ in Ω with $u = 0$ on $\partial\Omega$ would be bounded above by v. But obviously $u(x, y) = c \sin x \sin y$ is a solution, which can be made as large as one wishes by taking the constant $c > 0$ suitably large.

2.3 Maximum and sweeping principles for quasilinear inequalities

As a main consequence of the comparison Theorem 2.2.3 of the previous section we have the following

Theorem 2.3.1 (Maximum Principle). *Let* $v \in C^2(\Omega)$ *be a comparison function for* (2.2.1), *in the sense that there exists* M *such that*

(i) $v(x) \geq M$ *and* $Dv(x) \in \boldsymbol{P}$ *for all* $x \in \Omega$;

(ii) v *is an elliptic solution of the inequality*

$$a_{ij}(x, z, Dv(x)) \partial^2_{x_i x_j} v + B(x, z, Dv(x)) \leq 0 \qquad (2.3.1)$$

for all fixed values $z > M$.

If $u \in C^2(\Omega)$ *is a solution of* (2.2.1) *in* $U = \{x \in \Omega \, : \, Du(x) \in \boldsymbol{P}\}$ *and* $u \leq v$ *on* $\partial\Omega$, *then either* $u(x) \equiv v(x)$ *or* $u(x) < v(x)$ *in* Ω.

Proof. Define $\tilde{a}_{ij}(x, \boldsymbol{\xi}) = a_{ij}(x, u(x), \boldsymbol{\xi})$, $\tilde{B}(x, \boldsymbol{\xi}) = B(x, u(x), \boldsymbol{\xi})$ and

$$\mathscr{L}[v] = \tilde{a}_{ij}(x, Dv) \partial^2_{x_i x_j} v + \tilde{B}(x, Dv).$$

By (ii) the function v is an elliptic solution of the inequality $\mathscr{L}[v] \leq 0$ when $u(x) > M$. Moreover, obviously, $\mathscr{L}[u] \geq 0$ in U.

Let $\Omega' = \{x \in \Omega \, : \, u(x) > M\}$ and $U' = \{x \in \Omega' \, : \, Du(x) \in \boldsymbol{P}\} \subset U$. Since $u = M$ on $\partial\Omega' \cap \Omega$ and $u \leq v$ on $\partial\Omega$, it follows that $u \leq v$ on $\partial\Omega'$. Then by Theorem 2.2.3 applied to any component $\mathscr{C}'$ of Ω' we have $u \leq v$ in $\mathscr{C}'$, and so $u \leq v$ in Ω'. Hence $u \leq v$ in Ω.

The required conclusion now follows at once with the help of Theorem 2.2.1. $\qquad\square$

Theorem 2.3.1 is somewhat abstract, in that it depends on the *existence* of the comparison function v. When $[a_{ij}]$ and B are more specialized we can avoid this difficulty. In particular, consider the case where $\boldsymbol{Q} \subset \boldsymbol{B}_\varrho$ for some $\varrho \geq 0$ (the possibility $\boldsymbol{P} = \mathbb{R}^n$ is included when $\varrho = 0$). Assume that in $\Omega \times \mathbb{R}^+ \times \boldsymbol{P}$

$$\begin{cases} [a_{ij}(x, z, \boldsymbol{\xi})] \quad \text{is positive definite,} \\ B(x, z, \boldsymbol{\xi}) \leq \alpha|\boldsymbol{\xi}| \, E(x, z, \boldsymbol{\xi}) + \gamma, \\ \qquad \text{where } E(x, z, \boldsymbol{\xi}) := a_{ij}(x, z, \boldsymbol{\xi}) \xi_i \xi_j / |\boldsymbol{\xi}|^2, \end{cases} \qquad (2.3.2)$$

and α, γ are non-negative constants.

Theorem 2.3.2 (Maximum Principle). *Let $\boldsymbol{A}$ and B satisfy (2.3.2), and suppose that*

$$|\boldsymbol{\xi}|\, E(x, z, \boldsymbol{\xi}) \geq \Psi(|\boldsymbol{\xi}|) \qquad in \ \Omega \times \mathbb{R}^+ \times \boldsymbol{P}, \qquad \boldsymbol{P} = \mathbb{R}^n \setminus \boldsymbol{Q}, \quad (2.3.3)$$

where $\Psi = \Psi(t)$ is a strictly increasing function on (ϱ, ∞), $\varrho \geq 0$.

Let $u \in C^2(\Omega)$ be a solution of the boundary value problem

$$
\begin{aligned}
a_{ij}(x, u, Du)\partial^2_{x_i x_j} u + B(x, u, Du) &\geq 0 \qquad & in \ \Omega, \\
u &\leq 0 \qquad & on \ \partial\Omega,
\end{aligned}
\qquad (2.3.4)
$$

where $\Omega \subset \{x \in \mathbb{R}^n \ : \ 0 < x_1 < R\}$. Then there holds

$$u(x) \leq R \max\{\rho,\, C\}(e^k - 1), \qquad (2.3.5)$$

where[3]

$$
\begin{aligned}
C &= \Psi^{-1}(R\gamma), k = 1 + \alpha R, \qquad & when \ \lim_{t\to\infty} \Psi(t) > 2\gamma R, \\
C &= \Psi^{-1}(\ell), \quad k = 1 + (\alpha + \gamma/\ell)R, \\
& \qquad when \ \lim_{t\to\infty} \Psi(t) = 2\ell \leq 2\gamma R.
\end{aligned}
\qquad (2.3.6)
$$

For the important subcase of the p-Laplace operator one has $E(t) = (p-1)t^{p-2}$, $\Psi(t) = (p-1)t^{p-1}$ and $R\Psi^{-1}(R\gamma) = [\gamma/(p-1)]^{1/(p-1)}\, R^{p'}$.

Proof. It is enough to construct a comparison function $v = v(x)$ such that $v(x) > 0$ in Ω and (2.3.1) holds. Accordingly, we choose

$$v(x) = K(e^{mR} - e^{mx_1}), \qquad x \in \Omega,$$

where $m = k/R$, $K > R\max\{\varrho,\, C\}$. Then

$$\partial_{x_1} v(x) = -Kme^{mx_1}$$

---

[3]If $\Psi(\varrho) = \lim_{t\to\varrho+} \Psi(t) = \ell' > 0$ then we define $\Psi^{-1}(s) = \varrho$ when $s \leq \ell'$. Note that the case $\lim_{t\to\infty} \Psi(t) < \infty$ is possible. That is, take for $\boldsymbol{\xi} \neq \boldsymbol{0}$,

$$a_{ij}(\boldsymbol{\xi}) = \frac{2\ell}{|\boldsymbol{\xi}| + 1} \cdot \frac{\xi_i \xi_j}{|\boldsymbol{\xi}|^2};$$

an easy computation yields

$$E(\boldsymbol{\xi}) = \frac{2\ell}{|\boldsymbol{\xi}| + 1}, \qquad \Psi(t) = \frac{2\ell t}{t + 1}, \qquad \Psi^{-1}(s) = \frac{s}{2\ell - s},$$

so $\Psi(t) \to 2\ell$ as $t \to \infty$. (In fact in this case $\Psi^{-1}(\ell) = 1$.)

so $|Dv| \geq mK$ and $Dv \in \boldsymbol{P}$, since $m > 1/R$. Also

$$\partial^2_{x_1^2} v(x) = -Km^2 e^{mx_1} = -m|Dv|.$$

With the help of (2.3.2), a calculation shows that (2.3.1) is valid provided

$$m\,|Dv|\,a_{11}(x, z, Dv) \geq \alpha\,|Dv|\,E(x, z, Dv) + \gamma \qquad (2.3.7)$$

for all $x \in \Omega$ and $z > 0$. But $E(x, z, Dv) = a_{11}(x, z, Dv)$, so (2.3.7) becomes

$$m\,|Dv|\,E(x, z, Dv) \geq \alpha\,|Dv|\,E(x, z, Dv) + \gamma. \qquad (2.3.8)$$

Obviously (2.3.8) is satisfied if $(m - \alpha)\,|Dv|\,E(x, z, Dv) \geq \gamma$ for all $z > 0$. At the same time

$$|Dv|\,E(x, z, Dv) \geq \Psi(|Dv|) \geq \Psi(mK) \geq \Psi(C) \geq \min\{\gamma R, \ell\},$$

since $mK > (k/R)R\max\{\varrho, C\} \geq C$. Therefore (2.3.8) holds when k and C are given as in (2.3.6), and in turn (2.3.1) holds, as required.

We now apply Theorem 2.3.1, giving

$$u(x) \leq v(x) \leq K(e^k - 1) \qquad \text{in } \Omega.$$

Letting $K \to R\max\{\varrho, C\}$ completes the proof. $\qquad \square$

Remarks

1. The condition $u \leq 0$ on the boundary can obviously be replaced by $u \leq M$, by adding M to the right side of (2.3.5).
2. The condition $\Omega \subset \{x \in \mathbb{R}^n : 0 < x_1 < R\}$ can (by appropriate translation and rotation of coordinates) always be satisfied by any domain whose minimum diameter is R.
3. Finally, the theorem simplifies considerably when either $\boldsymbol{Q} = \emptyset$ or $\{0\}$ and Range $\Psi = \mathbb{R}^+$. Then $\varrho = 0$ and $u(x) \leq R\Psi^{-1}(\gamma R)\,[\exp(1 + \alpha R) - 1]$.
4. The possibility that $\boldsymbol{Q} \supsetneq \{0\}$, say $\boldsymbol{Q} = \boldsymbol{B}_\varrho$, $\varrho > 0$, is discussed later in Section 3.7.

The next result shows that when B is homogeneous the global condition (2.3.2) need be assumed only for $|\xi|$ small, clearly of importance in applications.

Theorem 2.3.3. *Assume $\boldsymbol{P} = \mathbb{R}^n$ or $\boldsymbol{P} = \mathbb{R}^n \setminus \{0\}$. Let the hypotheses of Theorem 2.3.2 hold, with the exceptions that $\gamma = 0$, and (2.3.2) and (2.3.3) are assumed to be valid only in $\Omega \times \mathbb{R}^+ \times \boldsymbol{R}_1$, $\boldsymbol{R}_1 = \{\boldsymbol{\xi} \in \mathbb{R}^n : 0 < |\boldsymbol{\xi}| < 1\}$. Let $u \in C^2(\Omega)$ be a solution of the boundary value problem (2.3.4) where Ω is now an arbitrary bounded domain in $\mathbb{R}^n$. Then $u \leq 0$ in Ω.*

In the generality of the present hypotheses, this seems to be a new result.

Proof. Since $\gamma = 0$ only the first case of (2.3.6) applies and so $C = \Psi^{-1}(0) = \varrho = 0$. In this case the constant $K > 0$ in the proof of Theorem 2.3.2 can be chosen arbitrarily small, and in particular so small that $|Dv(x)| \leq Kme^{mR} \leq 1$ in Ω. The rest of the proof of Theorem 2.3.2 then applies without change, giving $u \leq 0$ whatever the value of R.   □

Theorem 2.3.3 is false if one weakens condition (2.3.2), as follows from the example

$$\Delta_4 u + |Du|^2 = 0 \quad \text{in } B_R \subset \mathbb{R}^2. \tag{2.3.9}$$

Indeed, this equation has the solution $u(x) = \frac{1}{8}(R^2 - |x|^2)$ in B_R, which vanishes on the boundary, and at the same time is positive in the interior.

Theorem 2.3.4. *Let the hypotheses of Theorem 2.3.2 be satisfied, with the exception that (2.3.2) is replaced by the condition that*

$$B(x, z, \boldsymbol{\xi}) \leq (\alpha|\boldsymbol{\xi}| + \beta|\boldsymbol{\xi}|^q)E(x, z, \boldsymbol{\xi}) + \gamma, \qquad 0 < q < 1,$$

in $\Omega \times \mathbb{R}^+ \times \boldsymbol{P}$, where α, β, γ are non-negative constants.
Then (2.3.5) holds with the previous constant C replaced by $C + \beta^{1/(1-q)}$ and the previous constant k replaced by $k + 1$.

The proof is essentially the same as before. The additional term $\beta|\boldsymbol{\xi}|^q$ (in the case $q = 0$) was first introduced by Gilbarg and Trudinger ([43], Theorem 10.3).

The idea of Theorem 2.3.1 can be extended in the form of a *"field version"* of the result.

Theorem 2.3.5 (Sweeping Principle). *For $\lambda \in [0, 1]$, let $\lambda \mapsto v_\lambda = v(x, \lambda)$ be a family of $C^2(\Omega) \cap C(\overline{\Omega})$ functions which are strictly increasing in λ for each $x \in \overline{\Omega}$, and are such that v is of class $C(\overline{\Omega} \times [0, 1])$. Define*

$$\mathscr{L}[u](x) = a_{ij}(x, u, Du)\partial^2_{x_i x_i} u + B(x, u, Du), \qquad x \in \Omega. \tag{2.3.10}$$

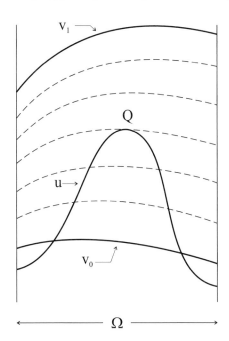

Figure 2.1: Proof of the Sweeping Principle: a contradiction with the Tangency Principle occurs at Q.

Assume that

$$\mathscr{L}[v_\lambda] \leq 0 \quad \text{in } \Omega, \qquad 0 \leq \lambda \leq 1,$$

that $\mathscr{L}[u] \geq 0$ and that $\mathscr{L}$ is elliptic either for u or for the family $\{v_\lambda\}_\lambda$. If $u \leq v_1$ in Ω and $u \leq v_0$ on $\partial\Omega$, then either $u \equiv v_0$ or $u < v_0$ in Ω.

The proof is an immediate consequence of Theorem 2.2.1, the idea being illustrated in the accompanying Figure 1, where the (solid line) function u is shown satisfying the conditions $u \leq v_1$ in Ω and $u \leq v_0$ on $\partial\Omega$, but at the same time contradicting the conclusion of the theorem. Note that in Theorem 2.3.5 no statement need be made concerning ellipticity of $\mathscr{L}$ along u or the monotonicity of $B(x, z, \boldsymbol{\xi})$ in z. The setting of Theorem 2.3.5 can be compared with the usual notion of a field used for sufficiency proofs in the calculus of variations.

The maximum principle Theorem 2.3.1 can be considered as essentially the special case $v_\lambda = v + C\lambda$ of the sweeping principle, C being chosen so that $u \leq v + C$ in Ω. A version of the sweeping principle in which the operator has a singular set $\boldsymbol{Q}$ can be left to the reader.

2.4   Comparison theorems for divergence structure inequalities

We consider the pair of differential inequalities

$$\operatorname{div} \boldsymbol{A}(x, Du) + B(x, u) \geq 0, \qquad\qquad (2.4.1)$$

$$\operatorname{div} \boldsymbol{A}(x, Dv) + B(x, v) \leq 0, \qquad\qquad (2.4.2)$$

in a bounded domain $\Omega \subset \mathbb{R}^n$. Let $\boldsymbol{A} : \Omega \times \mathbb{R}^n \to \mathbb{R}^n$ be in $L^\infty_{\mathrm{loc}}(\Omega \times \mathbb{R}^n)$, and $B : \Omega \times \mathbb{R} \to \mathbb{R}$ be in $L^\infty_{\mathrm{loc}}(\Omega \times \mathbb{R})$.

For the purpose of this section, by a *solution of* (2.4.1) *or* (2.4.2) *in* Ω we mean a (classical) distribution solution of class $C^1(\Omega)$, with the test function space consisting of all non-negative functions $\varphi \in C^1(\Omega)$ such that $\varphi \equiv 0$ near $\partial\Omega$. As is well known (see [43], Section 7.3) the test function space can without loss of generality be enlarged to include Lipschitz continuous functions which vanish near the boundary.

We shall treat here the simplest comparison theorems for divergence structure inequalities. More general results are given in Sections 3.4–3.6. Strong comparison theorems, under alternative hypotheses, have been obtained by Tolksdorf [108] and by Cuesta and Takáč [23].

Theorem 2.4.1 (Comparison Principle). *Let u and v be respective solutions of (2.4.1) and (2.4.2) in Ω. Suppose that $\boldsymbol{A} = \boldsymbol{A}(x, \boldsymbol{\xi})$ is independent of z and monotone in $\boldsymbol{\xi}$, i.e.,*

$$\langle \boldsymbol{A}(x, \boldsymbol{\xi}) - \boldsymbol{A}(x, \boldsymbol{\eta}), \boldsymbol{\xi} - \boldsymbol{\eta} \rangle > 0, \qquad when \ \boldsymbol{\xi} \neq \boldsymbol{\eta}; \qquad (2.4.3)$$

while $B = B(x, z)$ is independent of $\boldsymbol{\xi}$ and non-increasing in z.
    If $u \leq v$ on $\partial\Omega$, then $u \leq v$ in Ω.

Proof. Assume for contradiction that there exists $x_0 \in \Omega$ such that $u(x_0) > v(x_0)$. Let Γ be the open set $\{x \in \Omega : u(x) - v(x) > \varepsilon\}$, non-empty for $\varepsilon > 0$ sufficiently small. The function $\varphi = (u - v - \varepsilon)^+$ is uniformly Lipschitz continuous, has compact support in Ω, and $D\varphi = \boldsymbol{0}$ a.e. in $\Omega \setminus \Gamma$.

Subtracting (2.4.2) from (2.4.1) and using φ as test function yields

$$\int_\Gamma \langle \boldsymbol{A}(x, Du) - \boldsymbol{A}(x, Dv), Du - Dv \rangle$$

$$\leq \int_\Gamma [B(x, u) - B(x, v)](u - v - \varepsilon).$$

By (2.4.3) the left-hand side is positive unless $Du \equiv Dv$ in Γ, while the right-hand side is ≤ 0 since B is non-increasing in z. Let $\mathscr{C}$ be any component of Γ, so that $u - v \equiv \text{const.} = c$ in $\mathscr{C}$. If $\partial\mathscr{C} \cap \Omega \neq \emptyset$, then $c = \varepsilon$ which contradicts the fact that $u - v > \varepsilon$ in $\mathscr{C}$. Otherwise $\mathscr{C} = \Omega$ and c must be positive since $u(x_0) > v(x_0)$. This violates the fact that $u \leq v$ on $\partial\Omega$. $\square$

Remark. If B is non-increasing only for values $z \in (-\infty, \delta)$, then the conclusion of Theorem 2.4.1 continues to hold provided $u < \delta$ in Ω.

Because the condition (2.4.3) is somewhat abstract, it is of interest to exhibit explicit vector functions $\boldsymbol{A}$ for which (2.4.3) is satisfied. One of the simplest examples is

$$\boldsymbol{A} = \boldsymbol{A}(\boldsymbol{\xi}) = A(|\boldsymbol{\xi}|)\boldsymbol{\xi}, \quad \boldsymbol{\xi} \neq 0; \qquad \boldsymbol{A}(0) = 0, \tag{2.4.4}$$

where $s \mapsto A(s)$, $s > 0$, is positive and $\Phi(s) = sA(s)$ is strictly increasing on $\mathbb{R}^+$. We state this as

Proposition 2.4.2. *Let $\boldsymbol{\xi}$ and $\boldsymbol{\eta}$ be vectors in $\mathbb{R}^n$. Then for the function (2.4.4) we have*

$$\langle \boldsymbol{A}(\boldsymbol{\xi}) - \boldsymbol{A}(\boldsymbol{\eta}), \boldsymbol{\xi} - \boldsymbol{\eta} \rangle > 0$$

whenever $\boldsymbol{\xi} \neq \boldsymbol{\eta}$.

Proof. If one of the vectors is $\boldsymbol{0}$ the assertion is trivial. Otherwise, $\boldsymbol{\xi}, \boldsymbol{\eta} \neq \boldsymbol{0}$ and $\langle \boldsymbol{\xi}, \boldsymbol{\eta} \rangle \leq |\boldsymbol{\xi}| \cdot |\boldsymbol{\eta}|$, so that

$$\begin{aligned}
&\langle \boldsymbol{A}(\boldsymbol{\xi}) - \boldsymbol{A}(\boldsymbol{\eta}), \boldsymbol{\xi} - \boldsymbol{\eta} \rangle \\
&= A(|\boldsymbol{\xi}|)|\boldsymbol{\xi}|^2 + A(|\boldsymbol{\eta}|)|\boldsymbol{\eta}|^2 - A(|\boldsymbol{\xi}|)\langle \boldsymbol{\xi}, \boldsymbol{\eta} \rangle - A(|\boldsymbol{\eta}|)\langle \boldsymbol{\eta}, \boldsymbol{\xi} \rangle \\
&\geq \Phi(|\boldsymbol{\xi}|)|\boldsymbol{\xi}| + \Phi(|\boldsymbol{\eta}|)|\boldsymbol{\eta}| - \Phi(|\boldsymbol{\xi}|)|\boldsymbol{\eta}| - \Phi(|\boldsymbol{\eta}|)|\boldsymbol{\xi}| \\
&= \{\Phi(|\boldsymbol{\xi}|) - \Phi(|\boldsymbol{\eta}|)\}(|\boldsymbol{\xi}| - |\boldsymbol{\eta}|)
\end{aligned}$$

and the conclusion now comes from the strict monotonicity of Φ. $\square$

Proposition 2.4.2 obviously covers the p-Laplace operator $A(s) = s^{p-2}$, $p > 1$, as a special case. A second example of interest is the following

Proposition 2.4.3. *Suppose that $\boldsymbol{A}(x, \boldsymbol{\xi})$ is continuous in $\Omega \times \mathbb{R}^n$ and continuously differentiable with respect to $\boldsymbol{\xi}$ in the set $\Omega \times \mathbb{R}^n \setminus \{\boldsymbol{0}\}$, with the Jacobian matrix $[\partial_{\xi_j} A_i(x, \boldsymbol{\xi})]$ being positive definite.*
Then (2.4.3) is valid.

Proof. First we observe that if $\boldsymbol{\xi} \neq \boldsymbol{\eta}$ and the line segment $[\boldsymbol{\xi}, \boldsymbol{\eta}]$ does not include the point $\mathbf{0}$, then by the mean value theorem, for some point $\boldsymbol{\zeta}$ in the segment,

$$\langle \boldsymbol{A}(x, \boldsymbol{\xi}) - \boldsymbol{A}(x, \boldsymbol{\eta}), \boldsymbol{\xi} - \boldsymbol{\eta} \rangle = \langle \partial_{\boldsymbol{\xi}} \boldsymbol{A}(x, \boldsymbol{\zeta})(\boldsymbol{\xi} - \boldsymbol{\eta}), \boldsymbol{\xi} - \boldsymbol{\eta} \rangle > 0,$$

since $[\partial_{\boldsymbol{\xi}} \boldsymbol{A}(x, \boldsymbol{\xi})]$ is positive definite in $\Omega \times \mathbb{R}^n \setminus \{\mathbf{0}\}$.

When $\mathbf{0} \in [\boldsymbol{\xi}, \boldsymbol{\eta}]$, we apply the mean value theorem in each segment $[\boldsymbol{\xi}, \mathbf{0}]$, $[\mathbf{0}, \boldsymbol{\eta}]$, using the continuity of $\boldsymbol{A}$ with respect to $\boldsymbol{\xi}$ in $\Omega \times \mathbb{R}^n$. $\square$

Remark. In view of Proposition 2.4.3, when $\boldsymbol{A}$ is only continuous in the variable $\boldsymbol{\xi}$ the condition (2.4.3) is a generalization of the usual concept of ellipticity.

A delicate application of this proposition occurs when

$$\boldsymbol{A} = \boldsymbol{A}(x, \boldsymbol{\xi}) = A(|\boldsymbol{\xi}|)\boldsymbol{a}(x)\boldsymbol{\xi}, \qquad \boldsymbol{A}(x, \mathbf{0}) = \mathbf{0},$$

where $\boldsymbol{a} = \boldsymbol{a}(x) = [a_{ij}(x)]$, $i, j = 1, \ldots, n$, is a continuous real symmetric matrix defined in Ω, uniformly positive definite and satisfying

$$\lambda |\boldsymbol{\zeta}|^2 \leq a_{ij}(x) \zeta_i \zeta_j \leq \Lambda |\boldsymbol{\zeta}|^2, \qquad \lambda > 0, \tag{2.4.5}$$

for all $x \in \Omega$ and all $\boldsymbol{\zeta} \in \mathbb{R}^n$; we assume A has the properties noted before Proposition 2.4.2 and is continuously differentiable in $\mathbb{R}^+$.

Proposition 2.4.4. *Let $0 < \tau \leq \infty$ and assume*

$$\inf_{0 < s < \tau} \frac{sA'(s)}{A(s)} = c_1 > -1, \qquad \sup_{0 < s < \tau} \frac{sA'(s)}{A(s)} = c_2 < \infty, \tag{2.4.6}$$

and

$$\sqrt{\frac{\Lambda}{\lambda}} < \min\{\phi(c_1), \phi(c_2)\}, \qquad \phi(c) = \frac{2 + c + 2\sqrt{1 + c}}{|c|}. \tag{2.4.7}$$

Define $\boldsymbol{P} = \{\boldsymbol{\xi} \in \mathbb{R}^n : 0 < |\boldsymbol{\xi}| < \tau\}$. Then the operator $\boldsymbol{A}$ is elliptic in $\Omega \times \boldsymbol{P}$, in the sense that the Jacobian matrix $[\partial_{\boldsymbol{\xi}} \boldsymbol{A}(x, \boldsymbol{\xi})]$ is positive definite in $\Omega \times \boldsymbol{P}$. Moreover, for $x \in \Omega$,

$$\langle A(|\boldsymbol{\xi}|)\boldsymbol{a}(x)\boldsymbol{\xi} - A(|\boldsymbol{\eta}|)\boldsymbol{a}(x)\boldsymbol{\eta}, \boldsymbol{\xi} - \boldsymbol{\eta} \rangle > 0$$

whenever $\boldsymbol{\xi}, \boldsymbol{\eta} \in \overline{\boldsymbol{P}}$ and $\boldsymbol{\xi} \neq \boldsymbol{\eta}$.

The terms $\phi(c_1)$ or $\phi(c_2)$ respectively should be omitted from (2.4.7) if c_1 or $c_2 = 0$. If $c_1 = c_2 = 0$, then (2.4.7) itself should be omitted. Also, since $\phi(c) > 1$ for all $c > -1$ it is evident that (2.4.7) is automatically satisfied whenever $\boldsymbol{a}$ is a multiple of the identity.

Proof. We have $\partial_{\xi_j} A_i(x, \boldsymbol{\xi}) = A(|\boldsymbol{\xi}|) a_{ik}(x) b_{kj}(\boldsymbol{\xi})$ in $\Omega \times \boldsymbol{P}$, where

$$\boldsymbol{b} = [b_{ij}(\boldsymbol{\xi})] = \boldsymbol{I} + c \frac{\boldsymbol{\xi} \otimes \boldsymbol{\xi}}{|\boldsymbol{\xi}|^2}, \qquad c = c(|\boldsymbol{\xi}|) = \frac{|\boldsymbol{\xi}| A'(|\boldsymbol{\xi}|)}{A(|\boldsymbol{\xi}|)}.$$

The eigenvalues of $\boldsymbol{b}$ are 1, with multiplicity $n - 1$, and $1 + c$. Then from the Nicholson–Strang theorem, see [20, Theorem 2.1], it follows that $\boldsymbol{a} \boldsymbol{b}$, the product of the real matrices $\boldsymbol{a}$ and $\boldsymbol{b}$, will be positive definite provided $c = 0$ or

$$\left(\sqrt{\frac{\Lambda}{\lambda}} - 1 \right) \left(\sqrt{1 + c} - 1 \right) < 2 \ \text{ if } \ c > 0,$$

$$\left(\sqrt{\frac{\Lambda}{\lambda}} - 1 \right) \left(\sqrt{\frac{1}{1 + c}} - 1 \right) < 2 \ \text{ if } \ c < 0.$$

This however reduces to

$$\sqrt{\frac{\Lambda}{\lambda}} < \frac{2 + c + 2\sqrt{1 + c}}{|c|} = \phi(c). \tag{2.4.8}$$

By (2.4.6) we have $c_1 \leq c \leq c_2$ so by the monotonicity properties of ϕ there holds $\phi(c) \geq \min\{\phi(c_1), \phi(c_2)\}$, see Fig. 5.1 on page 114. Therefore in view of (2.4.7) the condition (2.4.8) holds for all $\boldsymbol{\xi} \in \boldsymbol{P}$. Thus $[\partial_{\boldsymbol{\xi}} \boldsymbol{A}(x, \boldsymbol{\xi})]$ is positive definite in $\Omega \times \boldsymbol{P}$.

Application of Proposition 2.4.3 then completes the proof (replacing $\mathbb{R}^n \setminus \{0\}$ in the proposition by the more general set $\boldsymbol{P}$ causes no difficulty). $\qquad \square$

With the help of the abstract comparison Theorem 2.4.1, the preceding Propositions 2.4.2 and 2.4.4 give explicit comparison principles for operators of the type

$$\boldsymbol{A} = \boldsymbol{A}(\boldsymbol{\xi}) = A(|\boldsymbol{\xi}|)\boldsymbol{\xi}, \quad \boldsymbol{A} = \boldsymbol{A}(x, \boldsymbol{\xi}) = A(|\boldsymbol{\xi}|)\boldsymbol{a}(x)\boldsymbol{\xi}, \quad (\boldsymbol{A} = 0 \text{ at } \boldsymbol{\xi} = 0).$$

For the special p-Laplace case $A(s) = s^{p-2}$, $p > 1$, that is $\boldsymbol{A}(x, \boldsymbol{\xi}) = |\boldsymbol{\xi}|^{p-2} \boldsymbol{a}(x)\boldsymbol{\xi}$, we have $c_1 = c_2 = p - 2 > -1$ in (2.4.6), whence (2.4.8) takes the form (for $p \neq 2$)

$$\sqrt{\frac{\Lambda}{\lambda}} < \frac{p + 2\sqrt{p - 1}}{|p - 2|}.$$

2.5 Tangency theorems via Harnack's inequality

Tangency theorems for non-divergence inequalities also have counterparts in the divergence structure case. We begin by considering the singular differential inequality

$$\operatorname{div}\tilde{\boldsymbol{A}}(x, u, Du) + \tilde{B}(x, u, Du) \leq 0 \quad \text{in } \Omega, \qquad u \geq 0, \tag{2.5.1}$$

where $\tilde{\boldsymbol{A}}$ and $\tilde{B}$ are in $L^\infty_{\text{loc}}(\Omega)$ and have the following homogeneity and ellipticity properties for all $x \in \Omega$, $z \in \mathbb{R}^+$ and $\boldsymbol{\xi} \in \mathbb{R}^n$:

$$\begin{aligned}
\langle \tilde{\boldsymbol{A}}(x, z, \boldsymbol{\xi}), \boldsymbol{\xi} \rangle &\geq a_1 |\boldsymbol{\xi}|^p - a_2 z^p, \\
|\tilde{\boldsymbol{A}}(x, z, \boldsymbol{\xi})| &\leq a_3 |\boldsymbol{\xi}|^{p-1} + a_4 z^{p-1}, \\
\tilde{B}(x, z, \boldsymbol{\xi}) &\geq -b_1 |\boldsymbol{\xi}|^{p-1} - b_2 z^{p-1},
\end{aligned} \tag{2.5.2}$$

with $p > 1$; $a_1, a_3 > 0$; $a_2, a_4, b_1, b_2 \geq 0$ being constants (see [92], where these conditions apparently appear first). The p-Laplace operator $\tilde{\boldsymbol{A}}(\boldsymbol{\xi}) = |\boldsymbol{\xi}|^{p-2}\boldsymbol{\xi}$, $p > 1$, clearly obeys the first line of (2.5.2) with $a_1 = a_3 = 1$ and $a_2 = a_4 = 0$.

Trudinger [109, Theorem 1.2], closely using the ideas of [92], has observed that under these conditions the following beautiful weak Harnack inequality is valid for non-negative solutions $u \in C^1(\Omega)$ of (2.5.1); see in particular Theorem 7.1.2.

For any ball B_R such that $0 < R \leq 1$ and $B_{2R} \subset \Omega$, there holds

$$\|u\|_{q, B_R} \leq C|R|^{n/q} \inf_{B_R} u(x), \tag{2.5.3}$$

where C depends only on $p, n, q; a_1, a_2, a_3, a_4, b_1, b_2$, while $q \in (0, (p-1)n/(n-p))$ (or $q \in \mathbb{R}^+$ if $p \geq n$).

This theorem holds equally for non-negative solutions of (2.5.1) in $W^{1,p}_{\text{loc}}(\Omega) \cap C(\Omega)$. The case when (2.5.1) is a linear inequality ($p = 2$) is of course included in the result.

The Harnack inequality immediately implies the following Strong Maximum Principle.[4]

Theorem 2.5.1 (Strong Maximum Principle). *Assume that the conditions (2.5.2) are valid only for $x \in \Omega$, $0 < z \leq 1$ and $|\boldsymbol{\xi}| \leq 1$. Let $u \in C^1(\Omega)$ be a (non-negative) distribution solution of (2.5.1) in Ω.*

Then either $u \equiv 0$ in Ω or $u > 0$ in Ω.

---

[4]The special case $a_2 = a_4 = 0$ and $B = 0$ was noted by Granlund [44].

Proof. We first modify $\tilde{A}$ and $\tilde{B}$ for values $u \geq 1$ and $|\boldsymbol{\xi}| > 1$, so that the modified functions remain in $L_{\text{loc}}^{\infty}(\Omega)$ but now also satisfy (2.5.2) for the complete set of variables. Then, corresponding to any classical (non-negative) solution of (2.5.1) for which $u(y) = 0$, there is some neighborhood $\mathscr{N}$ of y where $u \leq 1$ and $|Du| \leq 1$.

Let B_{2R} be a ball centered at y, with R so small that B_{2R} is in $\mathscr{N}$. Then $\min_{B_R} u(x) = 0$. In turn $\|u\|_{q,B_R} = 0$ by (2.5.3). That is, $u = 0$ in B_R. The conclusion $u \equiv 0$ in Ω now follows from connectedness, see the argument in the proof of Theorem 2.1.1. $\square$

Corresponding to Theorem 2.2.1, it is natural to seek a tangency principle which applies to C^1 solutions of divergence structure inequalities. To this end, we consider the partial differential inequalities

$$\begin{aligned}
\text{div}\, \boldsymbol{A}(x, u, Du) + B(x, u, Du) &\geq 0 \quad \text{in } \Omega, \\
\text{div}\, \boldsymbol{A}(x, v, Dv) + B(x, v, Dv) &\leq 0 \quad \text{in } \Omega,
\end{aligned} \tag{2.5.4}$$

where $\boldsymbol{A}$ and B are, respectively, a given vector and a given scalar function. Specifically, we assume that

$$\boldsymbol{A}(x, z, \boldsymbol{\xi}) : \Omega \times \mathbb{R} \times \mathbb{R}^n \to \mathbb{R}^n$$

is continuous, and continuously differentiable in the variables z and $\boldsymbol{\xi}$; at the same time

$$B(x, z, \boldsymbol{\xi}) : \Omega \times \mathbb{R} \times \mathbb{R}^n \to \mathbb{R} \tag{2.5.5}$$

is locally Lipschitz continuous in $\boldsymbol{\xi}$ and locally lower Lipschitz continuous in z, that is,

$$B(x, v, \boldsymbol{\eta}) - B(x, u, \boldsymbol{\xi}) \geq -b_1 |\boldsymbol{\eta} - \boldsymbol{\xi}| - L(v - u), \qquad \text{when } v > u,$$

for any compact set of arguments. The principal result of [94] is now the following

Theorem 2.5.2 (Tangency Principle). *Let* $u = u(x)$ *and* $v = v(x)$ *be functions of class* $C^1(\Omega)$, *satisfying the respective differential inequalities* (2.5.4). *Suppose that* $u \leq v$ *in* Ω, *and that at least one of the matrices*

$$[\partial_{\boldsymbol{\xi}} \boldsymbol{A}(x, u, Du)] \qquad \text{or} \qquad [\partial_{\boldsymbol{\xi}} \boldsymbol{A}(x, v, Dv)] \tag{2.5.6}$$

is positive definite in Ω. *Then either* $u \equiv v$ *or else* $u < v$ *throughout* Ω.

In [3] Almgren has obtained a related result for variational problems under somewhat less smoothness of the integrand than required above. This generalization is paid for, however, by a weaker conclusion, namely, that either $u \equiv v$ in Ω or else the set of equality is at most *of capacity zero*. Moreover, his theorem applies only to extremals and not to differential inequalities as is the case here.

At the end of the section, we also discuss the corresponding case when the solutions u and v are strongly differentiable rather than of class C^1.

Proof of Theorem 2.5.2. Let Ω' denote the subset of Ω where $u = v$. Obviously Ω' is relatively closed with respect to Ω. To complete the proof of the theorem it is therefore enough to show that Ω' is open, for then it must either be empty or coincide with Ω, since Ω is a connected set.

Thus assume that Ω' is not empty, and let y be an arbitrary point in Ω'. Obviously $u = v$ and $Du = Dv$ at y. Let B_R denote the closed ball of radius R centered at y, with $R \in (0, 1]$ so small that B_{3R} is contained in Ω. By subtracting and using the definition of weak solution we obtain, for x in B_{3R},

$$\operatorname{div}\{\boldsymbol{A}(x, v, Dv) - \boldsymbol{A}(x, u, Du)\} + B(x, v, Dv) - B(x, u, Du) \leq 0 \qquad (2.5.7)$$

By assumption, we have in B_{3R}

$$\begin{aligned} |\boldsymbol{A}(x, v, Dv) - \boldsymbol{A}(x, u, Du)| &\leq a|Dw| + bw, \\ -\{B(x, v, Dv) - B(x, u, Du)\} &\leq b_1|Dw| + b_2 w, \end{aligned} \qquad (2.5.8)$$

where $w = v - u \geq 0$ and a, b are suitable constants depending only on the structure of $\boldsymbol{A}$ and B and on bounds for u, v, Du, Dv in B_{3R}. Let us assume that the matrix $[\partial_\xi \boldsymbol{A}(x, u, Du)]$ in (2.5.6) is positive definite in Ω (the other case is treated similarly).

By continuity, the least eigenvalue of $[\partial_\xi \boldsymbol{A}(x, u, Du)]$ in B_{3R} is then positive, say equal to λ. In turn, for some vector $\boldsymbol{\zeta}$ in the line segment joining Du and Dv we have

$$\begin{aligned} \langle \boldsymbol{A}(x, u, Dv) &- \boldsymbol{A}(x, u, Du), Dw \rangle \\ &= \langle \partial_\xi \boldsymbol{A}(x, u, \boldsymbol{\zeta})Dw, Dw \rangle \\ &= \langle \partial_\xi \boldsymbol{A}(x, u, Du)Dw, Dw \rangle + o(|Dw|^2) \\ &\geq \tfrac{1}{2}\lambda|Dw|^2, \end{aligned}$$

if R is taken even smaller if necessary (since $Dw = \boldsymbol{0}$ at y).

Again using the differentiability properties of $\boldsymbol{A}$, we have next, for $x \in B_{3R}$,

$$
\begin{aligned}
\langle \boldsymbol{A}(x, v, Dv) &- \boldsymbol{A}(x, u, Du), Dw \rangle \\
&= \langle \boldsymbol{A}(x, u, Dv) - \boldsymbol{A}(x, u, Du), Dw \rangle \\
&\quad + \langle \boldsymbol{A}(x, v, Dv) - \boldsymbol{A}(x, u, Dv), Dw \rangle \\
&\geq \frac{\lambda}{2} |Dw|^2 - bw|Dw| \geq \frac{\lambda}{4} |Dw|^2 - \frac{b^2}{\lambda} w^2
\end{aligned}
\tag{2.5.9}
$$

by the Cauchy inequality.

We are now in position to apply the Harnack inequality (2.5.3). In particular, let the non-negative function $w = v - u$ be considered as a solution of the differential inequality (2.5.7), which we can write in the form (2.5.1) with w replacing u. Then in view of (2.5.8) and (2.5.9) the hypotheses (2.5.2) are satisfied with $p = 2$. Consequently, since $w = 0$ at y, we obtain the inequality (take $q = 1$)

$$
\int_{B_{2R}} w \leq 0.
$$

Hence $w = 0$ in B_{2R}. Therefore Ω' is an open set, completing the proof. $\qquad \square$

If the continuity and differentiability hypotheses on $\boldsymbol{A}$ and B in Theorem 2.5.2 are strengthened to hold *uniformly* in their variables, one can obtain a result applying not only to C^1 solutions of (2.5.4) but even to solutions in $W^{1,2}_{\mathrm{loc}}(\Omega) \cap C(\Omega)$. Supposing also that at least one of the matrices (2.5.6) is *uniformly* positive definite in any compact subset of Ω, we have the following conclusion.

Theorem 2.5.3. *Let $u = u(x)$ and $v = v(x)$ be solutions of (2.5.4) in the class $W^{1,2}_{\mathrm{loc}}(\Omega) \cap C(\Omega)$. Suppose that $u \leq v$ in Ω. Then either $u \equiv v$ or else $u < v$ throughout Ω.*

2.6 Uniqueness of the Dirichlet problem

A first case concerns semilinear equations in $\mathbb{R}^n$:

$$
\begin{aligned}
Lu + f(x, u) &= h(x) && \text{in } \Omega, \\
u &= g(x) && \text{on } \partial\Omega,
\end{aligned}
\tag{2.6.1}
$$

where L is the elliptic operator in Section 2.1.

Theorem 2.6.1. *Suppose $z \mapsto f(x, z)/z$, $z > 0$, is a non-increasing function for each fixed $x \in \Omega$, and assume $h(x) \leq 0$, $g(x) > 0$. Then the Dirichlet problem (2.6.1) can have at most one positive solution.*

Proof. Since $h(x) \leq 0$, one sees that the function $[f(x, z) - h(x)]/z$ is non-increasing in z. Let u, v be two positive solutions of (2.6.1). Since $u \leq v$ on $\partial\Omega$, it follows from Theorem 2.2.4 that $u \leq v$ in Ω. Similarly $v \leq u$ in Ω, and the proof is done.   □

As observed earlier, the problem (2.6.1) may possibly have no positive solutions at all; one can only state that if there does exist a positive solution it is unique.

The structure built up in Section 2.4 has as a consequence several uniqueness theorems for C^1 solutions of the singular Dirichlet problem

$$\begin{aligned} \operatorname{div} \boldsymbol{A}(x, Du) + B(x, u) &= 0 && \text{in } \Omega, \\ u &= u_0 && \text{on } \partial\Omega, \end{aligned} \qquad (2.6.2)$$

where $u_0 \in C(\partial\Omega)$, Ω is a bounded domain of $\mathbb{R}^n$, and $\boldsymbol{A}$ and B are as in Section 2.4.

Theorem 2.6.2. *Let condition (2.4.3) hold and assume that B is non-increasing in z. Then problem (2.6.2) can have at most one $C^1(\Omega)$ solution.*

This is an immediate consequence of Theorem 2.4.1. The special cases

$$\boldsymbol{A} = \boldsymbol{A}(\boldsymbol{\xi}) = A(|\boldsymbol{\xi}|)\boldsymbol{\xi}, \quad \boldsymbol{A} = \boldsymbol{A}(x, \boldsymbol{\xi}) = A(|\boldsymbol{\xi}|)\boldsymbol{a}(x)\boldsymbol{\xi} \qquad (\boldsymbol{A} = \boldsymbol{0} \text{ at } \boldsymbol{\xi} = \boldsymbol{0}),$$

given in Section 2.4 are of particular interest. For example, for the p-Laplace operator, $p > 1$, one has the following conclusion.

Corollary 2.6.3. *Let $B = B(x, z)$ be non-increasing in z. Then the Dirichlet problem*

$$\begin{aligned} \Delta_p u + B(x, u) &= 0 && \text{in } \Omega, \\ u &= u_0 && \text{on } \partial\Omega, \end{aligned} \qquad (2.6.3)$$

where $u_0 \in C(\partial\Omega)$, has at most one $C^1(\Omega)$ solution.

Of equal interest is a corresponding uniqueness theorem for C^1 solutions of the (non-singular) mean curvature equation (1.1.12); the formal statement can be omitted. For the restricted class of C^2 solutions this result was already noted in Section 2.1 as a consequence of the uniqueness theorem for solutions of non-singular quasilinear equations.

2.7 The boundary point lemma

Hopf's tangency Theorem 2.2.1 does not apply when $u - v$ attains a maximum at a boundary point of Ω. The following boundary point theorem treats this case.

Theorem 2.7.1. *Let $u = u(x)$ and $v = v(x)$ be solutions of the inequalities (2.2.1) and (2.2.2) in Ω, of class $C^2(\Omega) \cap C^1(\overline{\Omega})$. Assume $P = \mathbb{R}^n$, so that in particular the coefficient matrix $[a_{ij}]$ is continuous, and continuously differentiable with respect to z and ξ in the set $K = \Omega \times \mathbb{R} \times \mathbb{R}^n$, while similarly the scalar term B is continuously differentiable with respect to ξ in K.*

Assume that at least one of the solutions u and v is elliptic, and that B is locally lower Lipschitz continuous in the variable z, as in Theorem 2.2.1.

If $u < v$ in Ω, and $v = u$ at some point y on the boundary of Ω admitting an internally tangent sphere, then

$$\partial_\nu u > \partial_\nu v \quad \text{at } y.$$

McNabb [58] has treated the fully nonlinear version of Theorem 2.7.1, though his assumptions, when reduced to the quasilinear case, are stronger than required here. The theorem as stated follows directly from the boundary point Theorem 2.8.4, after applying the differencing procedure of Theorem 2.2.1 to obtain an appropriate linear inequality for the function $u - v$.

When we turn to $C^1(\overline{\Omega})$ solutions u and v of (2.5.4), it is a surprising fact that the analog of Theorem 2.7.1 is no longer true. This is shown by the following example due to Gilbarg ([42], page 169). Consider the function

$$u = u(x, y) = xe^{-\sqrt{|\log 4/r|}}, \qquad r^2 = x^2 + y^2,$$

where $n = 2$ and Ω is the domain $(x - 1)^2 + y^2 = 1$ in the (x, y)-plane. This function is of class C^1 in the closure of Ω and satisfies there the linear elliptic equation $\operatorname{div} A(x, y, Du) = 0$, where

$$A(x, y, Du) = (a\partial_x u + b\partial_y u, \, b\partial_x u + c\partial_y u)$$

with continuous coefficients

$$a = \frac{1}{\mu} + \frac{\mu^2 - 1}{r^2 \mu} y^2, \qquad b = \frac{1 - \mu^2}{r^2 \mu} xy, \qquad c = \frac{1}{\mu} + \frac{\mu^2 - 1}{r^2 \mu} x^2,$$

and $\mu = 1 + (2\sqrt{|\log 4/r|})^{-1}$. Clearly $u > 0$ in Ω, but u and Du are zero at the origin, contradicting the conclusion of Theorem 2.7.1.

In spite of this negative result, there are nevertheless two related results, analogous to Theorem 2.7.1 but applying to C^1 solutions of the divergence structure inequalities (2.5.4).

In the first case, the boundary point lemma holds for $C^1(\overline{\Omega})$ solutions of (2.5.4) *when $A(x, \boldsymbol{\xi})$ is linear in $\boldsymbol{\xi}$ and continuously differentiable in x, and B satisfies condition* (2.5.5).[5] This is a consequence of Hopf's construction (Lemma 2.8.2), together with the comparison principle Theorem 2.4.1. The proof can be left to the reader. Whether the condition of linearity can be avoided is an open question.

For convenience in stating the second result, we shall say that a boundary point y of Ω admits *an internal cone condition* provided there exists a right circular cone V with height h and vertex y which is contained in Ω.

Theorem 2.7.2. *Let $u = u(x)$ and $v = v(x)$ be functions of class $C^1(\overline{\Omega})$, satisfying the respective inequalities (2.5.4). Suppose that $u < v$ in Ω, and that at least one of the solutions is elliptic in Ω.*

Assume finally that $u = v$ at some point y on the boundary of Ω admitting an internal cone condition. Then the zero of $u - v$ at y is of finite order.

Proof. Assume for contradiction that $u - v$ has a zero of infinite order at y. Then $Du = Dv$ at y and the estimates (2.5.8) and (2.5.9) hold in the associated cone V (we may, of course, suppose that $h > 0$ is suitably small).

We can therefore apply the Harnack inequality to the positive function $w = v - u$ in any ball contained in V. This being the case, let us consider in particular a sequence of balls $B(y, \varrho)$, each of which is internally tangent to V and whose successive centers $y = y_{\varrho_k}$ and radii $\varrho = \varrho_k$ are such that

$$B(y_k, \varrho_k/3) \subset B(y_{k+1}, 2\varrho_{k+1}/3), \qquad k = 0, 1, 2, \ldots.$$

If ϑ is the half-angular opening of V, it is easy to see that the successive radii and centers can be chosen to satisfy the relation

$$\frac{\varrho_{k+1}}{\varrho_k} = \frac{|y_{k+1}|}{|y_k|} = \frac{1 + (1/3)\sin\vartheta}{1 + (2/3)\sin\vartheta} = \kappa < 1,$$

so that the sequence $B(y_k, \varrho_k)$ converges to y (for convenience we assume that y is the origin).

---

[5] A particular case of interest is the model Poisson equation $\Delta u + f(u) = 0$ when $f(u)$ is a locally Lipschitz continuous function.

By Theorem 1.2 of [109], see Section 2.5, there exists a constant C such that

$$\varrho^{-n} \int_{B(y,2\varrho/3)} w \leq C \min_{B(y,\varrho/3)} w(x)$$

for any ball $B(y, \varrho)$ in the sequence. On the other hand, for the ball $B(y', \varrho')$ preceding $B(y, \varrho)$ in the sequence we have (since $w > 0$)

$$\min_{B(y',\varrho'/3)} w(x) \leq \frac{3^n}{\omega_n \varrho'^n} \int_{B(y',\varrho'/3)} w \leq \frac{3^n}{\omega_n \varrho^n} \int_{B(y,2\varrho/3)} w,$$

where ω_n denotes the volume of the unit ball in n dimensions. Combining the last two inequalities now yields

$$\min_{B(y,\varrho/3)} w(x) \geq L \min_{B(y',\varrho'/3)} w(x),$$

where $L = \omega_n / 3^n C$. If this relation is iterated backward to successively larger radii ϱ, we find easily that

$$\min_{B(y_k,\varrho_k/3)} w(x) \geq L^k \min_{B(y_0,\varrho_0/3)} w(x),$$

whence $w(y_k) \geq \mathrm{const.} \, L^k$ for some positive constant and all positive integers k.

Now, by assumption, w has a zero of infinite order at y. Hence for any integer m there exists a constant $c(m)$ such that

$$w(y_k) \leq c(m)|y_k|^m = c(m)|y_0|^m \kappa^{mk}.$$

By combining the preceding two inequalities we obtain

$$\mathrm{const.} \, L^k \leq c(m)|y_0|^m \kappa^{mk}.$$

Letting k tend to infinity there results finally

$$\kappa^m \geq L,$$

which is impossible for sufficiently large m, since $\kappa < 1$. This completes the proof. $\qquad\square$

It is evident from the proof that one could determine an upper bound for the order of the zero at y depending on the structure of the coefficients A and B near the solution $u(x)$, namely, $m < \log L / |\log \kappa|$. We also note that an alternate proof of Theorem 2.7.2, in the case when equality holds in both relations (2.5.4), can be given on the basis of a result of Widman [115], though the proof as a whole would then be considerably more involved.

If the hypotheses on A and B are strengthened as in the last part of Section 2.6, then we can drop the condition that u and v are of class C^1. Specifically, in this case the following result holds.

Theorem 2.7.3. *Let $u = u(x)$ and $v = v(x)$ be continuous functions in the closure of $\overline{\Omega}$, possessing strong derivatives of class $L^2_{\mathrm{loc}}(\Omega)$. Suppose that $u \leq v$ in Ω and that (2.5.4) holds. Assume finally that $u = v$ at some point y on the boundary of Ω, admitting an internal cone condition. Then either $u \equiv v$ or else $u < v$ in Ω and the zero of $u - v$ at y is of finite order.*

Proof. Since (2.5.8) and (2.5.9) are valid in the present circumstances (see the demonstration of Theorem 2.5.3), the result follows exactly as in the proof of Theorem 2.7.2. □

2.8 Appendix: Proof of Eberhard Hopf's maximum principle

We begin with a simple but striking consequence of elementary calculus.

Theorem 2.8.1 (Weak Maximum Principle). *Let $u = u(x)$ be a C^2 function which satisfies the differential inequality*

$$Lu = \sum_{i,j} a_{ij}(x)\partial^2_{x_i x_j} u + \sum_i b_i(x)\partial_{x_i} u > 0$$

in a domain Ω, where the (symmetric) matrix $[a_{ij}]$ is positive semi-definite in Ω, but otherwise the coefficients a_{ij}, b_i are merely defined and finite at each point of Ω.

Then u cannot achieve an (interior) maximum in Ω. In particular, if $u \leq M$ on $\partial\Omega$, then $u \leq M$ in Ω.

Proof. If u reached the maximum value M at a point $y \in \Omega$, then since Ω is open we would have $Du(y) = \mathbf{0}$, while by elementary calculus the Hessian matrix $[\partial^2_{x_i x_j} u(y)]$ would be negative semi-definite, so that

$$\sum_{i,j} a_{ij}(y)\partial^2_{x_i x_j} u(y) \leq 0,$$

i.e., $Lu(y) \leq 0$, a contradiction. □

Lemma 2.8.2. *Let B_R be an arbitrary open ball of radius R in the domain Ω. Suppose that the (symmetric) matrix $[a_{ij}] = [a_{ij}(x)]$ is uniformly positive definite in B_R and the coefficients a_{ij}, $b_i = b_i(x)$ are uniformly bounded in B_R. Then for every constant $m > 0$ there exists a function $v \in C^2(\overline{B_R})$ such that*

(i) $v = 0$ *on* ∂B_R;
(ii) $v = m$ *on* $\partial B_{R/2}$;
(iii) $\partial_\nu v < 0$ *on* ∂B_R, *where ν is the exterior unit normal to B_R;*
(iv) $Lv > 0$ *in* $B_R \setminus \overline{B_{R/2}}$.

Proof. For a constant exponent $\alpha > 0$ still to be determined, we define

$$\tilde{v}(x) = e^{-\alpha r^2} - e^{-\alpha R^2}, \qquad x \in B_R, \tag{2.8.1}$$

where r denotes the distance from x to the center of B_R. Then

$$L\tilde{v}(x) = e^{-\alpha r^2}\Big\{ \Big(4\alpha^2 \sum_{i,j} a_{ij}(x)x_i x_j - 2\alpha \sum_i [a_{ii}(x) + b_i(x)x_i]\Big\},$$

where for simplicity we have taken the center of B_R as the origin $\mathbf{0}$ and $r = |x|$. Since by hypothesis $\sum_{i,j} a_{ij}(x)x_i x_j \geq \lambda r^2$, the constant α can be chosen so large that $L\tilde{v}(x) > 0$ for all x with $r = |x| \geq R/2$. Thus conditions (i), (iii) and (iv) hold for $\tilde{v}$. Define $v(x) = m\tilde{v}(x)/\tilde{v}(R/2)$, $x \in B_R$. Then v satisfies (ii) and of course continues to verify (i), (iii) and (iv). $\qquad\square$

Theorem 2.8.3 (Hopf's Boundary Point Lemma). *Suppose that the (symmetric) matrix $[a_{ij}] = [a_{ij}(x)]$ is uniformly positive definite in the domain Ω and that the coefficients a_{ij}, $b_i = b_i(x)$ are uniformly bounded in Ω. Let $u \in C^2(\Omega)$ satisfy the differential inequality $Lu \geq 0$ in Ω and let $x_0 \in \partial\Omega$ be such that*

(i) u *is continuous at x_0 and $\partial_\nu u$ exists at x_0, where ν is the outer normal vector to Ω at x_0;*
(ii) $u(x) < u(x_0)$ *for all $x \in \Omega$;*
(iii) *there exists a ball $B_R \subset \Omega$, with $x_0 \in \partial B_R$ (interior sphere condition).*
Then $\partial_\nu u(x_0) > 0$.

Proof. Let $u(x_0) = M$ and $\ell = \sup_{|x|=R/2} u(x) < M$. The function $w = u + v - M$ then satisfies $Lw > 0$ in $B_R \setminus \overline{B_{R/2}}$, while also $w \leq 0$ on ∂B_R and $\partial B_{R/2}$, provided $m = M - \ell$.

Consequently $w \leq 0$ in $B_R \setminus \overline{B_{R/2}}$ by Theorem 2.8.1, so that $\partial_\nu w(x_0) \geq 0$. In turn $\partial_\nu u(x_0) \geq -\partial_\nu v(x_0) > 0$. $\qquad\square$

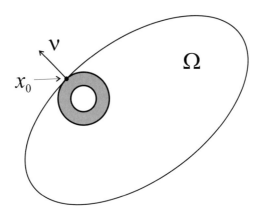

Figure 2.2: Proof of the Boundary Point Lemma; the annular region $B_R \setminus \overline{B_{R/2}}$ is shaded.

Proof of Hopf's Maximum Principle. Suppose u takes a maximum value M in Ω. The subset Ω_0 of Ω where $u = M$ is then non-empty and relatively closed in Ω. We must show that $\Omega_0 = \Omega$.

Thus suppose for contradiction that $\Omega_0 \neq \Omega$. By the connectedness of Ω it follows that the set $\partial\Omega_0 \cap \Omega$ must be non-empty (otherwise Ω_0 would be open as well as closed, and thus identical to Ω).

Fix $x_1 \in \partial\Omega_0 \cap \Omega$, and in turn let $\mathbf{0}$ be a point of Ω, as near to x_1 as we like, such that $u(\mathbf{0}) < M$. Taking $\mathbf{0}$ nearer to x_1 than to $\partial\Omega$, it follows that there is a largest open ball B_R in $\mathbb{R}^n$, with center at $\mathbf{0}$, which does not intersect Ω_0. Moreover $\overline{B}_R \subset \Omega$, so that in particular $u < M$ in B_R and $u = M$ at some point x_0 on the boundary of both B_R and Ω_0.

But then $\partial_\nu u(x_0) > 0$ by the boundary point Theorem 2.8.3. At the same time, x_0 is an *interior* maximum point of u; hence $Du(x_0) = \mathbf{0}$, an immediate contradiction. Thus $\Omega_0 = \Omega$, completing the proof. $\square$

The function $\tilde{v}$ in (2.8.1) was introduced by Hopf in [46]. An elegant alternative to $\tilde{v}$ is

$$\hat{v}(x) = r^{-\alpha} - R^{-\alpha}, \qquad \alpha > 0.$$

In fact

$$L\hat{v}(x) = \alpha r^{-\alpha} \Big\{ (\alpha + 2) \sum_{i,j} a_{ij}(x) x_i x_j - \sum_i [a_{ii}(x) + b_i(x) x_i] \, r^2 \Big\},$$

which is clearly positive in $B_R \setminus \overline{B_{R/2}}$ for suitably large α, as required.

The techniques used for the proof of the boundary point lemma yield another result of interest.

Theorem 2.8.4. *Let the hypotheses of Theorem 2.8.3 hold, with the exception that*

(a) *the inequality $Lu \geq 0$ is replaced by $[L + c(x)]u \geq 0$, where c is bounded below in a neighborhood of x_0, and*

(b) *either $u(x_0) = 0$ or $u(x_0) > 0$ and $c(x) \leq 0$. Then*

$$\partial_\nu u(x_0) > 0.$$

Proof. Consider first the case when $u(x_0) = 0$. Let d be a positive constant. From the proof of Lemma 2.8.2 it is easy to see that if the constant α is chosen even larger if necessary, then the function v given in Lemma 2.8.2 can equally be supposed to satisfy

(iv)′ $(L - d)v > 0$ in $E_R \equiv B_R \setminus \overline{B}_{R/2}$.

In turn $L(u + v) > -cu + dv$ in E_R.

As in the proof of Theorem 2.8.3, put $\ell = \sup_{|x|=R/2} u(x) < 0$. We claim that $u + v \leq 0$ in E_R. In fact, obviously $u + v \leq 0$ on $\partial B_R \cup \partial B_{R/2} = \partial E_R$, provided that $m = -\ell$. If the claim was false, there would be a point $y \in E_R$ at which $u + v$ would attain a positive maximum. Then we would have

$$L(u + v) > -(c + d)u > 0 \qquad \text{at } y, \qquad (2.8.2)$$

provided d is chosen so that $\inf_{x \in E_R} c(x) + d > 0$ (recall that c is bounded below in a neighborhood of x_0 and $u < 0$ in Ω). On the other hand, as in the proof of the weak maximum principle Theorem 2.8.1, we have necessarily $L(u + v) \leq 0$ at y, a contradiction with (2.8.2).

Thus $u + v \leq 0$ in E_R and in turn, since $u + v = 0$ at x_0, we obtain $\partial_\nu u(x_0) \geq -\partial_\nu v(x_0) > 0$, as required.

When $u(x_0) = M > 0$ we define $w = u - M$. Then $w(x_0) = 0$ and $[L + c(x)]w \geq -Mc(x) \geq 0$. The previous argument therefore yields $\partial_\nu u(x_0) = \partial_\nu w(x_0) > 0$. □

Corollary 2.8.5. *Let the hypotheses of Theorem 2.8.4 hold, with the exception that in condition (ii) of Theorem 2.8.3 one assumes only that $u(x) \leq u(x_0)$ for $x \in \Omega$. Then either*

$$u \equiv u(x_0) \quad \text{in } \Omega, \qquad \text{or} \qquad \partial_\nu u(x_0) > 0.$$

Proof. By Theorems 2.1.1 and 2.1.2, if $u \leq u(x_0)$ in Ω then either $u \equiv u(x_0)$ or $u < u(x_0)$ in Ω. The conclusion then follows from Theorem 2.8.4. □

Notes

The results in Section 2.1 are due to Eberhard Hopf. They are stated, however, in greater generality and in more usual notation. Theorems 2.2.1 and 2.2.3 are variants of Hopf's results; they are, however, new in the form given. Theorem 2.2.4 is also new here, using however an ingenious idea of Picone [71]. The maximum principle Theorem 2.3.2 corresponds to Theorem 10.3 of [43], though again formulated for the case of singular inequalities.

For maximum principles when u is not of class C^2, and even possibly only measurable, see, e.g., Littman [56] and Chapter 2 of Fraenkel [38]. For the case of distribution solutions, see Sections 2.4, 2.5 and Chapter 3.

The results of Section 2.4 are for the most part new, especially Proposition 2.4.4. The tangency principle Theorem 2.5.2 is due to Serrin [94]. The uniqueness Theorem 2.6.1 seems to be new. The proofs in Section 2.7 also follow those of [94].

The proof of the Hopf maximum principle in Section 2.8 is a streamlined version of that in [43]. The boundary point lemma, Theorem 2.8.3, appears first in [47]; see also Oleinik [68].

When the matrix $[a_{ij}]$ is semidefinite rather than positive definite, many of Hopf's results remain valid in appropriately modified and weakened forms, see [61]. Correspondingly, a weak maximum principle for parabolic equations or inequalities was given by Picone [72], and a strong maximum principle by Nirenberg [65]. These results are elegantly presented in the classical monograph of Protter and Weinberger [76, Chapter 3].

Problems

2.1 Show that the condition $c \leq 0$ is necessary in Theorem 2.1.1.

2.2 Show that the function $u(x) = -|x|^\alpha$, where $\alpha > 2$ is a real number, satisfies an equation of the form

$$\Delta u + c(x)u = 0, \qquad c(0) = -1,$$

with $c = c(x)$ discontinuous at $x = 0$, and negative and unbounded near $x = 0$.

Then show that the condition in Theorem 2.1.2 that the coefficient c be bounded below is necessary.

2.3 Prove the result stated after Theorem 2.1.2.
[*Hint.* Put $b_i(x) = -b(x)\partial_{x_i}u/|Du|$ when $Du(x) \neq \mathbf{0}$ and $b_i(x) = 0$ when $Du(x) = \mathbf{0}$. Then $b(x)|Du| = -\sum b_i(x)\partial_{x_i}u$.]

2.4 Supply the details for the proof of uniqueness for the Dirichlet problem for quasilinear equations stated after Theorem 2.1.5.

2.5 Suppose $\Delta u = -1$ in $B_R = \{x \in \mathbb{R}^n : |x| < R\}$, with $u = 0$ on ∂B_R. Find upper and lower bounds for u.

2.6 The function $u(x, y) = (1 - x^2 - y^2)/[(1 - x)^2 + y^2]$ is a solution of $\Delta u = 0$ in the unit disk B_1 of $\mathbb{R}^2$. Also $u = 0$ on ∂B_1 except at $(1, 0)$. Show that the maximum principle fails.

2.7 Show that a solution of $\Delta u = u^2$ in a domain Ω of $\mathbb{R}^n$ cannot attain a maximum in Ω unless $u \equiv 0$.

2.8 Show that the problem $\Delta u = -1$ in the two-dimensional square $Q = \{(x, y) \in \mathbb{R}^2 : |x|, |y| < 1\}$, with $u = 0$ on $|x| = 1$ and $\partial_x u - \partial_y u = 0$ on $|y| = 1$ has at most one solution. (Protter and Weinberger, [76].)

2.9 Let Ω be the square $Q_1 = \{(x, y) \in \mathbb{R}^2 : |x|, |y| < \pi/2\}$. Show that the Dirichlet problem for $\Delta u + u = 0$ in Q_1 has at most one solution. What about the square $\{(x, y) \in \mathbb{R}^2 : |x| < \pi/\sqrt{2}, |y| = \pi/\sqrt{2}\}$? [*Hint*: Use Theorem 2.3.2.]

2.10 Show that the divergence structure equation (2.3.9) can be written in the form

$$a_{ij}\partial^2_{x_i x_j}u + |Du|^2 = 0,$$

where $a_{ij} = |Du|^2\delta_{ij} + 2\partial_{x_i}u\partial_{x_j}u$. Thus show that $E(\xi) \geq |\xi|^2$ for equation (2.3.9), and hence that condition (2.3.2) fails for small values of ξ.

2.11 Find an estimate for $\sup_\Omega u(x)$ in Theorem 2.3.2 when the coefficient matrix $[a_{ij}]$ is a function only of the variable x.

In the next three problems, let $Lu = a_{ij}(x)\partial^2_{x_i x_j}u + b_i(x)\partial_{x_i}u$ be the linear differential operator in Hopf's main theorem, where the coefficients satisfy the conditions in Hopf's theorem.

2.12 If $Lu + c(x)u = f(x)$ in Ω, with $c < 0$ in Ω, show that

$$|u(x)| \leq \max_{\partial\Omega}|u| + \sup_\Omega |f/c| \qquad \text{in } \Omega.$$

This estimate is essentially due to Picone [71].

2.13 Let u, $v \in C^2(\Omega)$ be solutions of the differential inequalities

$$Lu + f(x, u) \geq 0, \qquad Lv + g(x, v) \leq 0$$

in Ω, with $v > 0$ in $\overline{\Omega}$.

Suppose that for each fixed $x \in \Omega$ there holds

$$\frac{g(x, t)}{t} - \frac{f(x, s)}{s} \geq 0, \qquad \text{when } s > t > 0.$$

If $u \leq v$ on $\partial\Omega$, then $u \leq v$ in Ω.

2.14 Let Ω satisfy an interior sphere condition at each point of $\partial\Omega$. Suppose that $u \in C^2(\Omega) \cap C^1(\overline{\Omega})$ satisfies

$$Lu = 0 \qquad \text{in } \Omega$$

and the boundary conditions

$$\alpha(x)u + \beta_i(x)\partial_{x_i}u = 0 \qquad \text{on } \partial\Omega,$$

where $\alpha(x) \cdot \langle \boldsymbol{\beta}(x), \boldsymbol{\nu} \rangle > 0$, and $\boldsymbol{\nu}$ is the exterior normal to $\partial\Omega$. Then $u \equiv 0$.

2.15 Let u, $v \in C^1(\overline{\Omega})$ be solutions of the differential inequalities (2.5.4), where $\boldsymbol{A} = \boldsymbol{A}(x, z, \boldsymbol{\xi})$ and $B = B(x, z, \boldsymbol{\xi})$ are continuously differentiable in the variables z and $\boldsymbol{\xi}$ in $\overline{\Omega} \times \mathbb{R} \times \mathbb{R}^n$. Suppose also that the $(n+1) \times (n+1)$ matrix

$$\begin{bmatrix} \partial_{\xi_1}\boldsymbol{A} & \cdots & \partial_{\xi_n}\boldsymbol{A} & \partial_z\boldsymbol{A} \\ -\partial_{\xi_1}B & \cdots & -\partial_{\xi_n}B & -\partial_z B \end{bmatrix} := \begin{bmatrix} \partial_{\xi_1}A_1 & \cdots & \partial_{\xi_n}A_1 & \partial_z A_1 \\ \vdots & \ddots & \vdots & \vdots \\ \partial_{\xi_1}A_n & \cdots & \partial_{\xi_n}A_n & \partial_z A_n \\ -\partial_{\xi_1}B & \cdots & -\partial_{\xi_n}B & -\partial_z B \end{bmatrix}$$

is non-negative definite in $\Omega \times \mathbb{R} \times \mathbb{R}^n$.

Show that if $u \leq v$ on $\partial\Omega$, then $u \leq v$ in Ω. (See [93], Theorem 6 on page 429).

2.16 Let $u \in C^2(\Omega) \cap C(\overline{\Omega})$ be a solution of $Lu + c(x)u = f(x)$ in a bounded C^1 domain Ω of $\mathbb{R}^n$, satisfying an exterior sphere condition at $x_0 \in \partial\Omega$, with $\overline{B}_R(y) \cap \overline{\Omega} = \{x_0\}$. Suppose $c \leq 0$ in Ω and let λ, Λ be positive constants such that for all $x \in \Omega$ and $\boldsymbol{\xi} \in \mathbb{R}^n$,

$$a_{ij}\xi_i\xi_j \geq \lambda|\boldsymbol{\xi}|^2 \quad \text{and} \quad |a_{ij}|, |b_i|, |c| \leq \Lambda.$$

If $\varphi \in C^2(\overline{\Omega})$ and $u = \varphi$ on $\partial\Omega$, show that u satisfies a Lipschitz condition at x_0,

$$|u(x) - u(x_0)| \le K|x - x_0| \qquad \text{for all } x \in \Omega,$$

where $K = K(\lambda, \Lambda, R, \operatorname{diam}\Omega, \sup_{\Omega}|f|, \|\varphi\|_{2,\Omega})$. Hence conclude that K provides a gradient bound for u on $\partial\Omega$, when $u \in C^1(\overline{\Omega})$ and $\partial\Omega$ is sufficiently smooth. If the sign of c is unrestricted, show that the same result holds provided K depends also on $\sup_{\Omega}|u|$. (Cf. [41, Problem 3.6].)

2.17 (Phragmèn–Lindelöf) Let u satisfy the inequality $\Delta u \ge 0$ in a sector Ω of angle opening π/α. Assume that $u \le M$ on the boundary $\vartheta = \pm\pi/2\alpha$ and that

$$\liminf_{R \to \infty} \{R^{-\alpha} \max_{r=R} u(r, \vartheta)\} \le 0.$$

Prove that $u \le M$ in Ω. (Cf. [76, Theorem 1.8 on page 94].)

2.18 Suppose that u satisfies the equation $\Delta u = 0$ in a domain Ω of $\mathbb{R}^n$, $n \ge 3$, except at a point $x_0 \in \Omega$. If

$$\lim_{r \to 0} r^{n-2} u(x) = 0, \qquad r = |x - x_0|,$$

show that u may be defined at x_0 so that $\Delta u = 0$ in Ω.
[*Hint*: The Dirichlet problem $\Delta u = 0$ in B, where B is an open ball of $\mathbb{R}^n$, has a unique solution given by the Poisson integral formula.]

Chapter 3

Maximum Principles for Divergence Structure Elliptic Differential Inequalities

3.1 Distribution solutions

For a large number of divergence structure equations, including equations which involve the important p-Laplacian operator Δ_p, there is a further series of maximum principles. In particular, in this chapter we study the differential inequality

$$\operatorname{div} \boldsymbol{A}(x, u, Du) + B(x, u, Du) \geq 0 \qquad \text{in } \Omega, \qquad (3.1.1)$$

where Ω is a bounded domain in $\mathbb{R}^n$ (unless otherwise stated explicitly), and

$$\boldsymbol{A}(x, z, \boldsymbol{\xi}) : \Omega \times \mathbb{R} \times \mathbb{R}^n \to \mathbb{R}^n, \qquad B(x, z, \boldsymbol{\xi}) : \Omega \times \mathbb{R} \times \mathbb{R}^n \to \mathbb{R}.$$

Throughout the chapter, by a *solution u of* (3.1.1) *in* Ω we mean specifically a distribution or weak solution, in the sense that $u \in L^1_{\mathrm{loc}}(\Omega)$ is weakly differentiable in Ω (that is, all its weak derivatives of first order exist); $\boldsymbol{A}(\,\cdot\,, u, Du)$, $B(\,\cdot\,, u, Du) \in L^1_{\mathrm{loc}}(\Omega)$; and

$$\int_{\Omega} \langle \boldsymbol{A}(x, u, Du), D\varphi \rangle \leq \int_{\Omega} B(x, u, Du)\varphi \qquad (3.1.2)$$

for all $\varphi \in C^1(\Omega)$ such that $\varphi \geq 0$ in Ω and $\varphi \equiv 0$ near $\partial\Omega$.

In order to treat solutions in the natural Sobolev space $W^{1,p}(\Omega)$ we shall require several preliminary results. We say that u is a *p-regular solution*, $p \geq 1$, if also[1]

$$\boldsymbol{A}(\,\cdot\,, u, Du) \in L_{\mathrm{loc}}^{p'}(\Omega), \qquad p' = p/(p-1). \tag{3.1.4}$$

Furthermore by $u \leq M$ on $\partial\Omega$ for some $M \in \mathbb{R}$ we mean explicitly that for every $\delta > 0$ there is a neighborhood of $\partial\Omega$ in which $u \leq M + \delta$.

For simplicity in printing, we shall write $\|\cdot\|_{\nu,\Gamma}$ for $\|\cdot\|_{L^\nu(\Gamma)}$ when Γ is a measurable subset of Ω, and $\|\cdot\|_\nu$ for $\|\cdot\|_{L^\nu(\Omega)}$.

Lemma 3.1.1. *Let f_h be the regularization (mollification) of a function $f \in L^p(\Omega)$, $p \geq 1$, with mollification radius h.*

Then $\|f_h - f\|_p \to 0$ as $h \to 0$; also a subsequence of $(f_h)_h$, which by agreement we identify as $(f_h)_h$, converges a.e. in Ω. Moreover if $f \in L^1(\Omega)$ and f is weakly differentiable in Ω, then in any domain $\Omega' \subset\subset \Omega$ we have $Df_h = [Df]_h$ for h sufficiently small.

For the proof of this lemma we refer to Lemmas 7.2 and 7.3 of [43].

Lemma 3.1.2. *Let $\psi : \mathbb{R} \to \mathbb{R}_0^+$ be a non-decreasing continuous function such that $\psi(t) = 0$ for $t \in (-\infty, \ell]$ and $\psi \in C^1$ for $t \in [\ell, \infty)$, with a possible corner at $t = \ell$ and with ψ' uniformly bounded. Let $u \in W_{\mathrm{loc}}^{1,p}(\Omega)$ be a p-regular solution of (3.1.1), and suppose that $f \in W_{\mathrm{loc}}^{1,p}(\Omega)$ is such that $f \leq \ell' < \ell$ on $\partial\Omega$.*

Then (3.1.2) is valid for $\varphi = \psi(f)$, in the sense that

$$\int_\Omega \langle \boldsymbol{A}(x, u, Du), D\varphi \rangle \leq \int_\Omega [B(x, u, Du)]^+ \varphi, \tag{3.1.5}$$

where $D\varphi = \psi'(f)Df$ when $f \neq \ell$.

Proof. The last line is a consequence of [43, Lemma 7.8].

---

[1]Condition (3.1.4) is obviously satisfied when $u \in W^{1,p}(\Omega)$ under the *"natural"* additional condition that, for all $(x, z, \boldsymbol{\xi})$ in $\Omega \times \mathbb{R} \times \mathbb{R}^n$,

$$|\boldsymbol{A}(x, z, \boldsymbol{\xi})| \leq a_3 |\boldsymbol{\xi}|^{p-1} + a_4 |z|^{p-1} + a_5, \tag{3.1.3}$$

where a_3, a_4 are constants and $a_5 \in L_{\mathrm{loc}}^{p'}(\Omega)$. The condition of p-regularity was noted in [92].

The principal requirement that $\boldsymbol{A}(\,\cdot\,, u, Du), B(\,\cdot\,, u, Du) \in L_{\mathrm{loc}}^1(\Omega)$ can be met if for example one assumes, in addition to (3.1.3), a corresponding condition on $B(x, z, \boldsymbol{\xi})$ and that both $\boldsymbol{A}(\,\cdot\,, u, Du), B(\,\cdot\,, u, Du)$ are measurable.

Let $\varphi_N = \psi_N(f)$ be the truncation of $\psi(f)$ at the level $N > \ell$, that is, equal to $\psi(f)$ when $f < N$ and to $\psi(N)$ when $f \geq N$. By the properties of ψ and the fact that $f < \ell'$ on $\partial\Omega$, it is clear that $\varphi_N \in W^{1,p}(\Omega)$ with $\varphi_N \equiv 0$ near $\partial\Omega$.

The regularization $\varphi_{N,h}$ of φ_N is in $C^1(\Omega)$ and vanishes near $\partial\Omega$ for h sufficiently small, and of course also $\varphi_{N,h} \geq 0$. Thus $\varphi_{N,h}$ can serve as a test function for (3.1.2), that is, by (3.1.2),

$$\int_\Omega \langle \boldsymbol{A}(x,u,Du), D\varphi_{N,h}\rangle \leq \int_\Omega [B(x,u,Du)]^+\varphi_{N,h}. \qquad (3.1.6)$$

By Lemma 3.1.1 we have $D\varphi_{N,h} = [D\varphi_N]_h$ for h sufficiently small; therefore

$$\|D\varphi_{N,h} - D\varphi_N\|_p \to 0, \qquad \varphi_{N,h} \to \varphi_N \qquad \text{a.e. in } \Omega \qquad (3.1.7)$$

as $h \to 0$. Clearly

$$A(\,\cdot\,,u,Du) \in L^{p'}_{\text{loc}}(\Omega) \quad \text{and} \quad [B(x,u,Du)]^+\varphi_{N,h} \leq N[B(x,u,Du)]^+$$

a.e. in Ω. Thus we can apply (3.1.7) to the left side of (3.1.6) and the dominated convergence theorem to the right side, since $[B(\,\cdot\,,u,Du)]^+ \in L^1_{\text{loc}}(\Omega)$. Hence for $h \to 0$ one gets

$$\int_\Omega \langle \boldsymbol{A}(x,u,Du), D\varphi_N\rangle \leq \int_\Omega [B(x,u,Du)]^+\varphi_N. \qquad (3.1.8)$$

Finally

$$\|D\varphi_N - D\varphi\|_p = \|D\varphi\|_{p,\{f\geq N\}} \to 0$$

as $N \to \infty$. Using the monotone convergence theorem (since $\varphi_N \nearrow \varphi$) proves the lemma. $\qquad\square$

The integral $\int B^+\varphi$ in (3.1.5) can at the moment possibly be infinite, though in our applications in the sequel it will in fact prove to be finite.

Lemma 3.1.3. *Lemma 3.1.2 applies to solutions $u \in W^{1,\infty}_{\text{loc}}(\Omega)$ with p-regularity no longer being required.*

The proof is essentially the same, with the exception that (3.1.7) is replaced by $D\varphi_{N,h} \to D\varphi_N$ a.e. in Ω as $h \to 0$, while by the definition of weak solution we have $A(\,\cdot\,,u,Du) \in L^1_{\text{loc}}(\Omega)$.

Appendix. The condition of p-regularity is necessary for the demonstration of Lemma 3.1.2. The delicacy of the structure can be emphasized by observing first that Gilbarg and Trudinger define weak solutions exactly as we do here (see equation (8.30) in [43]), while in their following Theorem 8.15 (for the case of linear equations) they consider solutions in $W^{1,2}(\Omega)$, these being 2-regular by linearity and so legitimate in forming test functions.

On the other hand, for Lemma 10.8 in [43, page 273] their solution is assumed to be in $C^1(\Omega)$, so one then must have $\boldsymbol{A}(\,\cdot\,,u,Du) \in L^1_{\mathrm{loc}}(\Omega)$ in order to use the theory of weak solutions. While not explicitly indicated in Lemma 10.8, this condition can be obtained from their earlier remark (page 260) that $\boldsymbol{A}$ is a differentiable function. But, once this is assumed, their structure condition (10.23) no longer applies, except when the exponent $p = 2$! There seems no way to avoid this dilemma other than giving up the differentiability of $\boldsymbol{A}$ and setting conditions so that $\boldsymbol{A}(\,\cdot\,,u,Du) \in L^1_{\mathrm{loc}}(\Omega)$, say that $\boldsymbol{A}$ is continuous in all its variables.

Even here, however, one must also deal with their later statement that solutions can be allowed in the space $W^{1,p}(\Omega)$, see [43, page 277]. This in turn requires the p-regularity condition $\boldsymbol{A}(\,\cdot\,,u,Du) \in L^{p'}_{\mathrm{loc}}(\Omega)$, a condition which is not indicated in [43].

Of course, this begs the question, under what conditions can one in fact obtain $\boldsymbol{A}(\,\cdot\,,u,Du) \in L^{p'}_{\mathrm{loc}}(\Omega)$ when $u \in W^{1,p}_{\mathrm{loc}}(\Omega)$? The simplest (though not the only) answer is found in the footnote above.

3.2   Maximum principles for homogeneous inequalities

Let the functions $\boldsymbol{A}$ and B in (3.1.1) be defined in the set $\Omega \times \mathbb{R}^+ \times \mathbb{R}^n$, and satisfy an alternative version of the natural p-homogeneous structure condition (2.5.2); that is, there are constants $a_1 > 0$ and $a_2, b_1, b_2 \geq 0$ such that for all $(x,z,\boldsymbol{\xi}) \in \Omega \times \mathbb{R}^+ \times \mathbb{R}^n$ there holds

$$\langle \boldsymbol{A}(x,z,\boldsymbol{\xi}),\boldsymbol{\xi}\rangle \geq a_1|\boldsymbol{\xi}|^p - a_2 z^p, \quad B(x,z,\boldsymbol{\xi}) \leq b_1|\boldsymbol{\xi}|^{p-1} + b_2 z^{p-1}, \quad (3.2.1)$$

where $p \in [1,\infty)$ describes the level of homogeneity of $\boldsymbol{A}$ and B. In particular, the case $p = 2$ covers linear elliptic inequalities of the form (3.1.1).

Theorem 3.2.1 (Maximum principle). *Assume $\boldsymbol{A}$ and B satisfy (3.2.1), with $a_2 = b_2 = 0$. Let $u \in W^{1,p}_{\mathrm{loc}}(\Omega)$, $p \geq 1$, be a p-regular solution of (3.1.1). If $u \leq M$ on $\partial\Omega$ for some constant $M \geq 0$, then $u \leq M$ a.e. in Ω.*

Proof. Since $a_2 = b_2 = 0$ it is enough to consider the case $M = 0$. Thus assume for contradiction that $\operatorname{essup}_\Omega u > 0$.

Let $\operatorname{essup}_\Omega u = V$. For $V < \infty$ fix $\ell \in (V/2, V)$ and for $V = \infty$ take $\ell > 1$. Define $\psi(t) = (t - \ell)^+$ and, as in Lemma 3.1.2, take $\varphi = \psi(u)$ as a (non-negative) test function for the inequality (3.1.1). Let

$$\Gamma' = \{x \in \Omega : \ell < u(x)\}.$$

Then since $\varphi = 0$, $D\varphi = \mathbf{0}$ a.e. in $\Omega \setminus \Gamma'$ and $D\varphi = Du$ in Γ, we see from (3.1.5) that

$$\int_{\Gamma'} \langle A(x, u, Du), Du \rangle \leq \int_{\Gamma'} [B(x, u, Du)]^+ \varphi.$$

Observing that $u > 0$ at all points where $\varphi > 0$, we can apply (3.2.1) with $a_2 = b_2 = 0$ to get

$$a_1 \int_{\Gamma'} |Du|^p \leq b_1 \int_{\Gamma'} |Du|^{p-1} \cdot \varphi. \tag{3.2.2}$$

Introduce the further set

$$\Gamma = \{x \in \Omega : \ell < u(x) < V\}. \tag{3.2.3}$$

We assert that (3.2.2) holds equally with the integration set Γ' replaced by Γ. If $V = \infty$ this is trivial. On the other hand if $V < \infty$ then

$$\int_{\Gamma'} |Du|^{p-1} \cdot \varphi = \left(\int_\Gamma + \int_{\{u=V\}} + \int_{\{u>V\}} \right) |Du|^{p-1} \cdot \varphi = \int_\Gamma |Du|^{p-1} \cdot \varphi$$

since $Du = \mathbf{0}$ a.e. where $u = V$ ([43], Lemma 7.7) while the set where $u > V$ has measure zero. Of course, in the same way $\|Du\|_{p,\Gamma'} = \|Du\|_{p,\Gamma}$, so the assertion is proved.

Restricting for the moment to the case $n > 1$, we put

$$s = \frac{n}{n-1} p.$$

Then replacing Γ' by Γ in (3.2.2) and applying Hölder's inequality to the right side yields

$$a_1 \|Du\|_{p,\Gamma}^p \leq b_1 |\Gamma|^{1/np} \|\varphi\|_{s,\Gamma} \|Du\|_{p,\Gamma}^{p-1}. \tag{3.2.4}$$

We claim that $\|Du\|_{p,\Gamma} > 0$. Indeed by Poincaré's inequality (Theorem 3.9.4)

$$\|\varphi\|_{p,\Omega} \leq Q\|D\varphi\|_{p,\Omega} = Q\|Du\|_{p,\Gamma'} = Q\|Du\|_{p,\Gamma}.$$

But $\|\varphi\|_{p,\Omega} = \|u\|_{p,\Gamma'} > 0$ since $\ell < \operatorname{essup}_\Omega u$, proving the claim. Now dividing (3.2.4) by $\|Du\|_{p,\Gamma}^{p-1}$ gives

$$a_1\|Du\|_{p,\Gamma} \leq b_1|\Gamma|^{1/np}\|\varphi\|_{s,\Gamma}. \tag{3.2.5}$$

Because φ vanishes near $\partial\Omega$, we have by Sobolev's inequality in the form given in Theorem 3.9.2,

$$\|\varphi\|_s \leq C\|D\varphi\|_p = C\|Du\|_{p,\Gamma} \leq (b_1/a_1)C|\Gamma|^{1/np}\|\varphi\|_s, \tag{3.2.6}$$

where $C = S(s_*, n)|\Omega|^{1/np'}$ and $s_* = n^2 p/(n^2 - n + np) \leq p$. Dividing by $\|\varphi\|_s \, (> 0)$ gives finally

$$a_1 \leq C|\Gamma|^{1/np} b_1. \tag{3.2.7}$$

But $\Gamma \to \emptyset$ as $\ell \to V$ by (3.2.3). This contradicts (3.2.7) and completes the proof for $n > 1$.

When $n = 1$ we set $s = \infty$. The proof is then unchanged except that the exponent $1/np$ becomes $1/p$, while (3.2.6) is replaced by

$$\|\varphi\|_\infty \leq \tfrac{1}{2}\|D\varphi\|_1 \leq \tfrac{1}{2}\|Du\|_p|\Omega|^{1/p'} \leq (b_1/2a_1)|\Omega|^{1/p'}|\Gamma|^{1/p}\|\varphi\|_\infty.$$

In turn $a_1 \leq C|\Gamma|^{1/p}b_1$, and the conclusion follows as before.　□

Theorem 3.2.2 (Maximum principle). *Assume that $\boldsymbol{A}$ and B satisfy (3.2.1) with $b_1 = b_2 = 0$. Let $u \in W^{1,p}_{\mathrm{loc}}(\Omega)$, $p > 1$, be a p-regular solution of the inequality (3.1.1) in Ω. If $u \leq 0$ on $\partial\Omega$, then $u \leq 0$ a.e. in Ω.*

Proof. Assume for contradiction that $V = \operatorname{essup}_\Omega u > 0$, possibly infinite. For $\ell > 0$ define $\psi(t) = 0$ when $t \leq \ell$ and

$$\psi(t) = 1 - (\ell/t)^{p-1}$$

for $t \geq \ell$. Lemma 3.1.2 now applies, so that $\varphi = \psi(u)$ can be used as a (non-negative) test function for (3.1.1). That is, by (3.1.5) with $b_1 = b_2 = 0$,

$$\int_\Gamma \langle \boldsymbol{A}(x, u, Du), D\varphi \rangle \leq 0, \tag{3.2.8}$$

where $\Gamma = \{x \in \Omega \; : \; u(x) > \ell\}$. Using the relations

$$D\varphi = \psi'(u)Du, \qquad \psi'(u) = (p-1)\ell^{p-1}u^{-p} \qquad \text{a.e. in } \Gamma,$$

we obtain from (3.2.8) and (3.2.1), after dividing by $(p-1)\ell^{p-1}$,

$$0 \geq \int_\Gamma \frac{\langle A(x,u,Du), Du \rangle}{u^p} \geq \int_\Gamma \frac{a_1|Du|^p - a_2 u^p}{u^p}, \qquad (3.2.9)$$

that is

$$a_1 \int_\Gamma |D \log u|^p \leq a_2 |\Gamma|. \qquad (3.2.10)$$

Define $\varphi_1(x) = \log(u(x)/\ell)$ if $u(x) > \ell$ and $\varphi_1(x) = 0$ if $u(x) \leq \ell$. As in the proof of Lemma 3.1.2 it is clear that φ_1 is in $W^{1,p}(\Omega)$. Moreover, since $\varphi_1 = 0$ in $\Omega \setminus \Gamma$, it then follows from Sobolev's inequality (Theorem 3.9.2) that

$$\|\varphi_1\|_{s,\Omega} \leq C \|D\varphi_1\|_{p,\Omega} = C \left\|D \log \frac{u}{\ell}\right\|_{p,\Gamma} = C \|D \log u\|_{p,\Gamma}, \; (3.2.11)$$

where, as before, $s = p^*$ if $p < n$ and $s \in (p, \infty)$ if $p \geq n$.

Now take $\ell \leq \min\{1, V/2\}$, and define

$$\Sigma = \begin{cases} \{x \in \Omega \; : \; V/2 \leq u(x) \leq V\}, & \text{when } V < \infty, \\ \{x \in \Omega \; : \; u(x) \geq 1\}, & \text{when } V = \infty. \end{cases}$$

In the first case, since $\varphi_1 \geq \log(V/2\ell)$ in Σ, we find from (3.2.10) and (3.2.11) that

$$|\Sigma|^{1/s} \log \frac{V}{2\ell} \leq C \left(\frac{a_2}{a_1}|\Gamma|\right)^{1/p},$$

which gives a contradiction as $\ell \to 0$ (since Σ is independent of ℓ and $|\Gamma| \leq |\Omega|$). In the second case, similarly, since $\varphi_1 \geq \log(1/\ell)$ in Σ,

$$|\Sigma|^{1/s} \log \frac{1}{\ell} \leq C \left(\frac{a_2}{a_1}|\Gamma|\right)^{1/p},$$

and again there is a contradiction as $\ell \to 0$. $\qquad\qquad\qquad\qquad\qquad\square$

Remarks

1. An alternative formulation of the boundary condition requires that $(u - M)^+ \in W_0^{1,p}(\Omega)$. In this case, (3.1.4) must be strengthened to

$$\boldsymbol{A}(\cdot, u, Du), \ B(\cdot u, Du) \in L^{p'}(\Omega),$$

   and corresponding changes are needed for the following proofs.

2. It is obvious that condition (3.2.1) in the previous theorems needs to be valid only for the range of values $u(x)$, $Du(x)$, $x \in \Omega$. We shall take advantage of this remark in later sections where it is assumed that $u \in W_{\mathrm{loc}}^{1,\infty}(\Omega)$ rather than $u \in W_{\mathrm{loc}}^{1,p}(\Omega)$.

3. If Ω is unbounded and the boundary condition is understood to include the limit relation

$$\limsup_{|x|\to\infty, \ x\in\Omega} u(x) \le M, \tag{3.2.12}$$

   then the conclusions of Theorems 3.2.1 and 3.2.2 continue to hold.

Theorem 3.2.3. *In Theorem 3.2.1 the coefficient b_1 can be taken in an appropriate Lebesgue space, that is*

$$b_1 \in \begin{cases} L_{\mathrm{loc}}^{n/(1-\varepsilon)}(\Omega), & when \quad 1 \le p \le n, \\ L_{\mathrm{loc}}^{p}(\Omega), & when \quad\quad p > n, \end{cases} \tag{3.2.13}$$

for some $\varepsilon \in (0, 1]$.

   The same result holds for Theorem 3.2.2 when $a_2 \in L^1(\Omega)$ (for all $p > 1$).

Proof. When $1 < p \le n$ the proof of Theorem 3.2.1 is valid exactly as before, with (3.2.4) replaced by

$$a_1\|Du\|_{p,\Gamma}^p \le |\Gamma|^{\varepsilon/n} \, \|b_1\|_{n/(1-\varepsilon),\Gamma} \, \|u\|_{p^*,\Gamma} \, \|Du\|_{p,\Gamma}^{p-1}.$$

For the case $p \ge n$, see Theorems 6.1.4 and 6.1.5 below with $a = b = 0$.

   The second result is obvious from the proof as given. $\qquad\square$

Theorem 3.2.4. *The conclusions of Theorems 3.2.1 and 3.2.2 remain valid when the right side of (3.2.1) is replaced by*

$$\begin{aligned} \langle \boldsymbol{A}(x, z, \boldsymbol{\xi}), \boldsymbol{\xi}\rangle &\ge a_1|\boldsymbol{\xi}|^p - a_2 z^p, \\ B(x, z, \boldsymbol{\xi}) &\le b_1\big(|\boldsymbol{\xi}|^{p-1} + |\boldsymbol{\xi}|^{q-1}\big), \end{aligned} \tag{3.2.14}$$

with $1 < q < p$.

The proofs are essentially the same as before, except that (3.2.2), for example, now becomes

$$a_1 \int_\Gamma |Du|^p \le b_1 \left(\int_\Gamma \{|Du|^{p-1} + |Du|^{q-1}\} \cdot |w| \right).$$

One then applies Hölder's inequality to the separate terms on the right side, as before. The details may be left to the reader.

3.3 A maximum principle for thin sets

When the coefficients a_2, b_1, b_2 in (3.2.1) do not vanish, the maximum principles Theorems 3.2.1 and 3.2.2 are no longer valid, as one can see from obvious examples, e.g., the equation $\Delta u + u = 0$ in a ball, as in elementary eigenvalue theory.

Nevertheless, if the domain in question has sufficiently small measure, that is, is sufficiently "*thin*", then the maximum principle remains correct even when a_2, b_1, b_2 are non-zero.

Theorem 3.3.1 (Maximum principle). *Assume A and B satisfy (3.2.1), and let $u \in W^{1,p}_{loc}(\Omega)$, $p \ge 1$, be a p-regular solution of (3.1.1). Suppose also that the measure of Ω is so small that*

$$\left[\left(\frac{b_1}{a_1} \right)^p + p \frac{a_2 + b_2}{a_1} \right]^{n/p} |\Omega| < \omega_n, \qquad (3.3.1)$$

where ω_n is the measure of the unit ball in $\mathbb{R}^n$.

If $u \le 0$ on $\partial\Omega$, then $u \le 0$ a.e. in Ω.

Proof. Define $\varphi = (u - \varepsilon)^+$ for $\varepsilon > 0$. Then, as in the proof of Theorem 3.2.1, see (3.2.2),

$$\int_\Gamma (a_1 |Du|^p - a_2 u^p) \le \int_\Gamma (b_1 u |Du|^{p-1} + b_2 u^p),$$

where $\Gamma = \{x \in \Omega : u(x) > \varepsilon\}$. In turn, using the Hölder and Young inequalities, with $c = a_2 + b_2$,

$$\int_\Gamma |Du|^p \le \frac{b_1}{a_1} \|u\|_{p,\Gamma} \|Du\|_{p,\Gamma}^{p-1} + \frac{c}{a_1} \int_\Gamma u^p$$

$$\le \frac{1}{p'} \|Du\|_{p,\Gamma}^p + \frac{1}{p} \left(\frac{b_1}{a_1} \|u\|_{p,\Gamma} \right)^p + \frac{c}{a_1} \int_\Gamma u^p$$

(note $1/p' = 0$ when $p = 1$).

Hence

$$\|Du\|_{p,\Gamma} \leq \left[\left(\frac{b_1}{a_1}\right)^p + p\,\frac{c}{a_1}\right]^{1/p} \|u\|_{p,\Gamma}.$$

Next, by Poincaré's inequality (Theorem 3.9.4),

$$\|u\|_{p,\Gamma} \leq \|u - \varepsilon\|_{p,\Gamma} + \|\varepsilon\|_{p,\Gamma} = \|(u - \varepsilon)^+\|_{p,\Omega} + \varepsilon|\Gamma|^{1/p}$$

$$\leq \left(\frac{|\Omega|}{\omega_n}\right)^{1/n} \|Du\|_{p,\Gamma} + \varepsilon|\Omega|^{1/p},$$

since $D[(u - \varepsilon)^+] = \mathbf{0}$ a.e. in $\Omega \setminus \Gamma$ and $D[(u - \varepsilon)^+] = Du$ in Γ. Combining the previous two lines gives

$$\|u\|_{p,\Gamma} \leq \left(\frac{|\Omega|}{\omega_n}\right)^{1/n} \left[\left(\frac{b_1}{a_1}\right)^p + p\,\frac{c}{a_1}\right]^{1/p} \|u\|_{p,\Gamma} + \varepsilon|\Omega|^{1/p}$$

$$\leq (1 - \theta)\|u\|_{p,\Gamma} + \varepsilon|\Omega|^{1/p}$$

for some $\theta \in (0,1)$. Hence $\|u\|_{p,\Gamma} \leq (\varepsilon/\theta)|\Omega|^{1/p}$. Letting $\varepsilon \to 0$ and using the monotone convergence theorem then gives $\|u^+\|_p = 0$. Consequently $u \leq 0$ a.e. in Ω. □

Condition (3.2.1) includes the p-Laplace operator Δ_p. For this case the coefficient in (3.3.1) becomes

$$(b_1^p + p\,b_2)^{n/p};$$

in particular, $p = 2$ in the Laplace case. For a related but much deeper result, see Problem 6.5.

A real number λ such that the Dirichlet problem

$$\operatorname{div}\boldsymbol{A}(x, u, Du) + B(x, u, Du) + \lambda|u|^{p-2}u = 0 \quad \text{in } \Omega, \quad p > 1,$$
$$u = 0 \quad \text{on } \partial\Omega, \tag{3.3.2}$$

has a non-trivial solution is called an eigenvalue for (3.3.2). With the help of the thin set Theorem 3.3.1 one can give a lower estimate for any possible eigenvalue of (3.3.2). We state this as

Corollary 3.3.2. *Let $u \in W^{1,p}_{\mathrm{loc}}(\Omega)$, $p > 1$, be a non-trivial p-regular solution of (3.3.2). Assume $\boldsymbol{A}$ and B satisfy (3.2.1) in the stronger form*

$$\langle \boldsymbol{A}(x, z, \boldsymbol{\xi}), \boldsymbol{\xi}\rangle \geq a_1|\boldsymbol{\xi}|^p - a_2|z|^p, \qquad |B(x, z, \boldsymbol{\xi})| \leq b_1|\boldsymbol{\xi}|^{p-1} + b_2|z|^{p-1}.$$

Then

$$\lambda + a_2 + b_2 \geq \frac{a_1}{p}\left[\kappa^p - \left(\frac{b_1}{a_1}\right)^p\right], \qquad \kappa = \left(\frac{\omega_n}{|\Omega|}\right)^{1/n}.$$

The proof is left to the reader. In the canonical case $B = 0$ the corollary yields the estimate $\lambda + a_2 \geq a_1\kappa^p/p$ for the eigenvalues of the pure operator $\operatorname{div}\boldsymbol{A}(x, u, Du)$ with homogeneous Dirichlet data.

3.4 A comparison theorem in $W^{1,p}(\Omega)$

As in Section 2.4, consider the pair of differential inequalities (2.4.1) and (2.4.2), with $\boldsymbol{A}$ and B no longer required to be in $L^\infty_{\mathrm{loc}}(\Omega)$.

Theorem 3.4.1. *Let u and v be respectively p-regular solutions of (2.4.1) and (2.4.2) of class $W^{1,p}_{\mathrm{loc}}(\Omega)$. Suppose that $\boldsymbol{A} = \boldsymbol{A}(x, \boldsymbol{\xi})$ is independent of z and monotone in $\boldsymbol{\xi}$, i.e., (2.4.3) holds, while $B = B(x, z)$ is independent of $\boldsymbol{\xi}$ and non-increasing in z.*
If $u \leq v$ on $\partial\Omega$, then $u \leq v$ a.e. in Ω.

Proof. By definition of distribution solution we get by subtraction

$$\int_\Omega \langle \boldsymbol{A}(x, Du) - \boldsymbol{A}(x, Dv), D\varphi\rangle \leq \int_\Omega [B(x, u) - B(x, v)]\varphi.$$

Taking $\varphi = (u - v - \ell)^+$, $\ell > 0$, as test function, we find from Lemma 3.1.2 and (2.4.3) that

$$0 \leq \int_\Gamma \langle \boldsymbol{A}(x, Du) - \boldsymbol{A}(x, Dv), Du - Dv\rangle$$
$$\leq \int_\Omega [B(x, u) - B(x, v)]^+ (u - v - \ell)^+,$$

where $\Gamma = \{x \in \Omega : u - v - \ell > 0\}$. Since B is non-increasing in the variable z the right-hand side is zero. Hence $Du = Dv$ a.e. in Γ. Consequently, in view of [43, Lemma 7.6 (a)], we have

$$D\varphi = \begin{cases} Du - Dv & \text{in } \Gamma \\ 0 & \text{in } \Omega \setminus \Gamma \end{cases} = 0 \quad \text{a.e. in } \Omega.$$

That is, the function φ, considered as an element of $W^{1,p}(\Omega)$, has weak derivative zero and vanishes near $\partial\Omega$. Hence by the Poincaré inequality,

Theorem 3.9.4, there holds

$$\|\varphi\|_p \leq C\|D\varphi\|_p = 0.$$

Therefore $\varphi = (u - v - \ell)^+ = 0$ a.e. in Ω, that is, $u \leq v + \ell$ a.e. in Ω. Letting $\ell \to 0$ completes the proof. $\qquad\square$

The special case where $\boldsymbol{A}$ satisfies (2.4.4) is of particular interest, since it includes the p-Laplace operator $A(s) = s^{p-2}$, $p > 1$. That is, we have the following

Corollary 3.4.2. *Let $\boldsymbol{A}$ have the form (2.4.4), and suppose that $B = B(x, z)$ is independent of $\boldsymbol{\xi}$ and non-increasing in z. Let u and v be respectively p-regular solutions of (2.4.1) and (2.4.2) of class $W_{\mathrm{loc}}^{1,p}(\Omega)$. If $u \leq v$ on $\partial\Omega$, then $u \leq v$ in Ω.*

Proof. This is a direct consequence of Theorem 3.4.1 and Proposition 2.4.2. $\qquad\square$

For the case of the p-Laplace operator, it is clear that the p-regularity of any solution is automatic. Corollary 3.4.2 applies also to the mean curvature operator $A(s) = 1/\sqrt{1 + s^2}$. Here $|\boldsymbol{A}(\boldsymbol{\xi})| < 1$, so 1-regularity of a solution $u \in W_{\mathrm{loc}}^{1,1}(\Omega)$ is again automatic.

Proposition 2.4.3 can also be applied in the present case, though we can omit the details. Finally, the uniqueness theorems given in Section 2.6 obviously carry over to the present case. In particular, if $B = B(x, z)$ is non-increasing in z, the Dirichlet problem (2.6.3) when $u_0 \in C(\partial\Omega)$ has at most one solution in $W_{\mathrm{loc}}^{1,p}(\Omega)$. Similarly the mean curvature Dirichlet problem

$$\mathrm{div}\left(\frac{Du}{\sqrt{1 + |Du|^2}}\right) + B(x, u) = 0 \qquad \text{in } \Omega,$$
$$u = u_0 \qquad \text{on } \partial\Omega,$$

has at most one solution in $W_{\mathrm{loc}}^{1,1}(\Omega)$. This last result seems to be new.

3.5 Comparison theorems for singular elliptic inequalities

When the relatively simple assumptions of the previous section do not apply, in particular when the function B depends explicitly on the variable $\boldsymbol{\xi}$, or the operator $\boldsymbol{A}$ is singular at more than isolated points, one can nevertheless reach useful conclusions by applying the maximum principles of Section 3.2. These include the well-known results of Chapter 10 of [43], and in turn lead to the mostly new uniqueness theorems in the later Section 3.8.

Consider the pair of differential inequalities

$$\operatorname{div} \boldsymbol{A}(x, u, Du) + B(x, u, Du) \geq 0 \qquad \text{in } \Omega, \qquad (3.5.1)$$
$$\operatorname{div} \boldsymbol{A}(x, v, Dv) + B(x, v, Dv) \leq 0 \qquad \text{in } \Omega, \qquad (3.5.2)$$

where Ω is a bounded domain in $\mathbb{R}^n$, and

$$\boldsymbol{A} : \Omega \times \mathbb{R} \times \mathbb{R}^n \to \mathbb{R}^n, \qquad B : \Omega \times \mathbb{R} \times \mathbb{R}^n \to \mathbb{R}.$$

As in Section 3.1, by a *solution of* (3.5.2) *in* Ω we mean a distribution or weak solution, in the sense that $v \in L^1_{\text{loc}}(\Omega)$ is weakly differentiable in Ω, $\boldsymbol{A}(\,\cdot\,, v, Dv), B(\,\cdot\,, v, Dv) \in L^1_{\text{loc}}(\Omega)$ and

$$\int_\Omega \langle \boldsymbol{A}(x, v, Dv), D\varphi \rangle \geq \int_\Omega B(x, v, Dv)\varphi \qquad (3.5.3)$$

for all non-negative functions $\varphi \in C^1(\Omega)$ such that $\varphi \equiv 0$ near $\partial\Omega$.

Among other topics, we shall deal with singular or degenerate inequalities in which ellipticity disappears as $\boldsymbol{\xi} \to \boldsymbol{0}$, as for the p-Laplace operator Δ_p, $p \neq 2$, where $\boldsymbol{A} = \boldsymbol{A}_p(\boldsymbol{\xi}) = |\boldsymbol{\xi}|^{p-2}\boldsymbol{\xi}$. In fact, in many cases the arguments by which a singular point $\boldsymbol{0}$ is treated can be generalized to allow for larger singular sets. A structure in which such behavior can be studied is described in the following principal conditions, which we assume throughout this and the next two sections.

(i) $\boldsymbol{A}$ *is continuous with respect to* $\boldsymbol{\xi}$ *in* $\Omega \times \mathbb{R} \times \mathbb{R}^n$.

(ii) *There exists a non-empty open subset* $\boldsymbol{P}$ *of* $\mathbb{R}^n$ *(possibly* $\boldsymbol{P} = \mathbb{R}^n$) *such that* $\boldsymbol{A}$ *is continuously differentiable with respect to* $\boldsymbol{\xi}$ *in* $\Omega \times \mathbb{R} \times \boldsymbol{P}$.

$\boldsymbol{P}$ is called the *regular* set for the inequalities (3.5.1) and (3.5.2), while

$$\boldsymbol{Q} = \mathbb{R}^n \setminus \boldsymbol{P}$$

is the *singular* set. If $\boldsymbol{Q} = \emptyset$ the problem is called *regular*, while otherwise it is *singular*. We say that the operator $\boldsymbol{A}$ is (*strictly*) *elliptic* in a set $K \subset \Omega \times \mathbb{R} \times \boldsymbol{P}$ if the Jacobian matrix $[\partial_\xi \boldsymbol{A}]$ is (*uniformly*) *positive definite* in K.[2]

---

[2]The concept of strict ellipticity can be illustrated with the example of the p-Laplace operator, where $\boldsymbol{A}(\boldsymbol{\xi}) = \boldsymbol{A}_p(\boldsymbol{\xi}) = |\boldsymbol{\xi}|^{p-2}\boldsymbol{\xi}$. This is elliptic for $\boldsymbol{\xi} \neq \boldsymbol{0}$ when $p > 1$, but strictly elliptic only when $1 < p \leq 2$.

In stating our next results, it is convenient to define

$$\boldsymbol{B}_r = \{\boldsymbol{\xi} \in \mathbb{R}^n \ : \ |\boldsymbol{\xi}| \leq r\}, \qquad \boldsymbol{R}_r = \boldsymbol{B}_r \setminus \{\boldsymbol{0}\}.$$

The following comparison principle then holds, both for regular operators as well as singular operators for which the singular set is the single point $\boldsymbol{Q} = \{\boldsymbol{0}\}$.

Theorem 3.5.1 (Comparison Principle). *Let $\boldsymbol{Q} = \emptyset$ or $\boldsymbol{Q} = \{\boldsymbol{0}\}$. Suppose that $\boldsymbol{A} = \boldsymbol{A}(x, \boldsymbol{\xi})$ is independent of z and strictly elliptic in $\Omega \times \boldsymbol{R}_r$ for all $r > 0$. Assume additionally that $B(x, z, \boldsymbol{\xi})$ is locally Lipschitz continuous with respect to $\boldsymbol{\xi}$ in $\Omega \times \mathbb{R} \times \mathbb{R}^n$ and moreover is non-increasing in z.*

Let u and v be solutions of (3.5.1) and (3.5.2) of class $W_{\text{loc}}^{1,\infty}(\Omega)$ in Ω. If $u \leq v + M$ on $\partial\Omega$, where M is constant, then $u \leq v + M$ in Ω.

Proof. We treat only the case $\boldsymbol{Q} = \{\boldsymbol{0}\}$. When $\boldsymbol{Q}$ is empty the proof is slightly simpler, and can be omitted. Moreover, since $\boldsymbol{A}$ is independent of z it is enough to consider only the case $M = 0$.

Step 1. Suppose that $(x, \boldsymbol{\xi})$, $(x, \boldsymbol{\eta}) \in K' = \Omega \times \boldsymbol{R}_W$ for some $W > 0$. If $\boldsymbol{\xi} \neq \boldsymbol{\eta}$ and the line segment $[\boldsymbol{\xi}, \boldsymbol{\eta}]$ does not include the point $\boldsymbol{0}$, then by the mean value theorem, for some point $\boldsymbol{\zeta}$ in the segment,

$$\langle \boldsymbol{A}(x, \boldsymbol{\xi}) - \boldsymbol{A}(x, \boldsymbol{\eta}), \boldsymbol{\xi} - \boldsymbol{\eta} \rangle = \langle \partial_{\boldsymbol{\xi}} \boldsymbol{A}(x, \boldsymbol{\zeta})(\boldsymbol{\xi} - \boldsymbol{\eta}), \boldsymbol{\xi} - \boldsymbol{\eta} \rangle.$$

Since by hypothesis the matrix $[\partial_{\boldsymbol{\xi}} \boldsymbol{A}(x, \boldsymbol{\xi})]$ is uniformly positive definite in $\Omega \times \boldsymbol{R}_W$, it follows that

$$\langle \boldsymbol{A}(x, \boldsymbol{\xi}) - \boldsymbol{A}(x, \boldsymbol{\eta}), \boldsymbol{\xi} - \boldsymbol{\eta} \rangle \geq a_1 |\boldsymbol{\xi} - \boldsymbol{\eta}|^2, \tag{3.5.4}$$

where

$$a_1 = \inf_{x \in \Omega, \ \boldsymbol{\xi} \in \boldsymbol{R}_W} \{\text{min eigenvalue of } [\partial_{\boldsymbol{\xi}} \boldsymbol{A}(x, \boldsymbol{\xi})]\} > 0.$$

We claim that (3.5.4) holds also when $\boldsymbol{0} \in [\boldsymbol{\xi}, \boldsymbol{\eta}]$. First, if $\boldsymbol{0}$ is an end point of $[\boldsymbol{\xi}, \boldsymbol{\eta}]$, say $\boldsymbol{\eta} = \boldsymbol{0}$, it is enough to let $\boldsymbol{\eta} \to \boldsymbol{0}$ in (3.5.4), since $\boldsymbol{A}$ is continuous at $\boldsymbol{0}$ and a_1 remains unchanged. The remaining possibility, when $\boldsymbol{0}$ is in the interior of $[\boldsymbol{\xi}, \boldsymbol{\eta}]$ is now obvious.

Next, if $(x, u, \boldsymbol{\xi})$, $(x, v, \boldsymbol{\eta}) \in K''$, where K'' is a compact subset of $\Omega \times \mathbb{R} \times \mathbb{R}^n$, then by local Lipschitz continuity of B we have

$$B(x, u, \boldsymbol{\xi}) - B(x, v, \boldsymbol{\eta}) \leq b_1 |\boldsymbol{\xi} - \boldsymbol{\eta}| + B(x, u, \boldsymbol{\eta}) - B(x, v, \boldsymbol{\eta}),$$

where b_1 is the Lipschitz constant of B in the set K''. In particular, since B is non-increasing in z,

$$B(x, u, \boldsymbol{\xi}) - B(x, v, \boldsymbol{\eta}) \leq b_1 |\boldsymbol{\xi} - \boldsymbol{\eta}| \qquad \text{when } u > v. \qquad (3.5.5)$$

Step 2. By subtracting (3.5.1) and (3.5.2) we get

$$\operatorname{div}\{\boldsymbol{A}(x, Du) - \boldsymbol{A}(x, Dv)\} + B(x, u, Du) - B(x, v, Dv) \geq 0 \qquad (3.5.6)$$

in Ω. Let $w = u - v$ and define

$$\tilde{\boldsymbol{A}}(x, \boldsymbol{\xi}) = \boldsymbol{A}(x, \boldsymbol{\xi} + Dv(x)) - \boldsymbol{A}(x, Dv(x)).$$

Clearly

$$\tilde{\boldsymbol{A}}(x, Dw) = \boldsymbol{A}(x, Du) - \boldsymbol{A}(x, Dv),$$

so that in view of (3.5.6) the function w can be considered as a solution of the differential inequality

$$\operatorname{div} \tilde{\boldsymbol{A}}(x, Dw) + \tilde{B}(x, w, Dw) \geq 0 \qquad (3.5.7)$$

where $\tilde{B}(x, z, \boldsymbol{\xi}) = B(x, z + v(x), \boldsymbol{\xi} + Dv(x)) - B(x, v(x), Dv(x))$ is defined analogously to $\tilde{\boldsymbol{A}}$. Of course, also $w = u - v \leq 0$ on $\partial\Omega$.

Since $u, v \in W^{1,\infty}_{\text{loc}}(\Omega)$ it follows that in any compact subset Ω' of Ω we have $Du, Dv \in \boldsymbol{R}_W$ for some $W > 0$. Thus (3.5.4) and (3.5.5) hold in Ω' with the identifications $\boldsymbol{\xi} = Du$ and $\boldsymbol{\eta} = Dv$ (so $\boldsymbol{\xi} - \boldsymbol{\eta} = Dw$); that is we have

$$\langle \tilde{\boldsymbol{A}}(x, Dw), Dw \rangle \geq a_1 |Dw|^2$$

and

$$\tilde{B}(x, w, Dw) \leq b_1 |Dw| \qquad \text{when } w > 0.$$

Stated in other terms, the functions $\tilde{\boldsymbol{A}}$ and $\tilde{B}$ in (3.5.7) obey the structural conditions (3.2.1) *along the solution* w, that is, with $\boldsymbol{\xi} = Dw$ and with also $a_2 = b_2 = 0$, $p = 2$.

Since $w \leq 0$ on $\partial\Omega$ we can therefore apply Theorem 3.2.1 to obtain $w \leq 0$ in Ω, that is $u \leq v$. $\qquad \square$

Remarks. This is essentially Theorem 10.7 (i) of [43] with the important exceptions that $\boldsymbol{A}$ and B are allowed to be singular at $\boldsymbol{\xi} = \boldsymbol{0}$, and that the class $C^1(\Omega)$ is weakened to $W^{1,\infty}_{\text{loc}}(\Omega)$. Compare also Theorem 10.3 of [81].

If Ω is unbounded and the boundary condition is understood to include the limit relation

$$\limsup_{|x|\to\infty,\ x\in\Omega}\ \{u(x)-v(x)\}\le M,$$

then the conclusion of Theorem 3.5.1 continues to hold. The same conclusion is valid for the later results of the section.

In the important case of the p-Laplace operator (where $\boldsymbol{Q}=\{\boldsymbol{0}\}$) we have the following corollary of Theorem 3.5.1.

Corollary 3.5.2. *Let u and v be solutions in $W^{1,\infty}_{\mathrm{loc}}(\Omega)$ of the inequalities*

$$\Delta_p u + B(x,u,Du)\ge 0,\qquad \Delta_p v + B(x,v,Dv)\le 0 \quad \text{in}\ \ \Omega,$$

where $1<p\le 2$. Assume also that $B=B(x,z,\boldsymbol{\xi})$ is locally Lipschitz continuous with respect to $\boldsymbol{\xi}$ in $\Omega\times\mathbb{R}\times\mathbb{R}^n$ and is non-increasing in the variable z. If $u\le v+M$ on $\partial\Omega$, where M is constant, then $u\le v+M$ in Ω.

Proof. Here $\boldsymbol{A}(\boldsymbol{\xi})=|\boldsymbol{\xi}|^{p-2}\boldsymbol{\xi}$ (and $\boldsymbol{A}(\boldsymbol{0})=\boldsymbol{0}$), so by direct calculation

$$\partial_{\boldsymbol{\xi}}\boldsymbol{A}(\boldsymbol{\xi})=|\boldsymbol{\xi}|^{p-2}\left[\mathbb{I}_n+(p-2)\frac{\boldsymbol{\xi}\otimes\boldsymbol{\xi}}{|\boldsymbol{\xi}|^2}\right],\qquad \boldsymbol{\xi}\ne\boldsymbol{0}.$$

Therefore the minimum eigenvalue of the Jacobian matrix $[\partial_{\boldsymbol{\xi}}\boldsymbol{A}(\boldsymbol{\xi})]$ when $\boldsymbol{\xi}\ne\boldsymbol{0}$ is $(p-1)|\boldsymbol{\xi}|^{p-2}$ and so $a_1=(p-1)W^{p-2}$ in (3.5.4) since $p\le 2$. That is, $\boldsymbol{A}$ is strictly elliptic in $\mathbb{R}^n\setminus\{\boldsymbol{0}\}$. $\qquad\square$

Corollary 3.5.2 can be compared with the results of Section 2.4. In particular, by the final remarks there *the restriction $1<p\le 2$ is unnecessary when $B=B(x,z)$ is independent of the variable $\boldsymbol{\xi}$.* See also Corollary 3.6.3 below.

Theorem 3.5.3 (Comparison Principle). *Suppose that $\boldsymbol{A}$ is strictly elliptic in $\Omega\times B_r\times\boldsymbol{R}_r$ for every $r>0$, and $\partial_z\boldsymbol{A}$ is locally bounded in $\Omega\times\mathbb{R}\times\mathbb{R}^n$. Assume additionally that $B=B(x,z)$ does not depend on $\boldsymbol{\xi}$ and is non-increasing in the variable z.*

Let u and v be solutions of (3.5.1) and (3.5.2) of class $W^{1,\infty}_{\mathrm{loc}}(\Omega)$. If $u\le v$ on $\partial\Omega$, then $u\le v$ in Ω.

Proof. The proof is essentially the same as that for Theorem 3.5.1, with the exception that the difference expression in (3.5.4) is treated differently, that is

$$\langle \boldsymbol{A}(x,u,\boldsymbol{\xi}) - \boldsymbol{A}(x,v,\boldsymbol{\eta}), \boldsymbol{\xi} - \boldsymbol{\eta}\rangle \geq \langle \boldsymbol{A}(x,u,\boldsymbol{\xi}) - \boldsymbol{A}(x,u,\boldsymbol{\eta}), \boldsymbol{\xi} - \boldsymbol{\eta}\rangle$$
$$+ \langle \boldsymbol{A}(x,u,\boldsymbol{\eta}) - \boldsymbol{A}(x,v,\boldsymbol{\eta}), \boldsymbol{\xi} - \boldsymbol{\eta}\rangle$$
$$= I_1 + I_2.$$

Now $I_1 \geq a_1|\boldsymbol{\xi} - \boldsymbol{\eta}|^2$ as in (3.5.4). Also by the mean value theorem

$$I_2 = \langle \partial_z \boldsymbol{A}(x,t,\boldsymbol{\eta}), \boldsymbol{\xi} - \boldsymbol{\eta}\rangle(u-v) \geq -c_1|\boldsymbol{\xi} - \boldsymbol{\eta}| \cdot |u-v|,$$

where t is in the open interval between u and v, and $c_1 = \sup_{K'} |\partial_z \boldsymbol{A}(x,z,\boldsymbol{\xi})|$. By Cauchy's inequality this yields $I_2 \geq -a_1|\boldsymbol{\xi} - \boldsymbol{\eta}|^2/2 - 2c_1^2(u-v)^2/a_1$. In combination, in place of (3.5.4) we now have

$$\langle \boldsymbol{A}(x,u,\boldsymbol{\xi}) - \boldsymbol{A}(x,v,\boldsymbol{\eta}), \boldsymbol{\xi} - \boldsymbol{\eta}\rangle \geq \tfrac{1}{2}a_1|\boldsymbol{\xi} - \boldsymbol{\eta}|^2 - a_2(u-v)^2, \quad (3.5.8)$$

where $a_2 = 2c_1^2/a_1$. In addition

$$B(x,u) - B(x,v) \leq 0 \qquad \text{when } u > v. \qquad (3.5.9)$$

Now let $w = u - v$, $\boldsymbol{\xi} = Dw$ and define

$$\tilde{\boldsymbol{A}}(x,z,\boldsymbol{\xi}) = \boldsymbol{A}(x,z+v(x),\boldsymbol{\xi}+Dv(x)) - \boldsymbol{A}(x,v(x),Dv(x)).$$

Then proceeding as in the proof of Theorem 3.5.1 we obtain

$$\operatorname{div}\tilde{\boldsymbol{A}}(x,w,Dw) + \tilde{B}(x,w) \geq 0.$$

Similarly, from the fact that u, $v \in W^{1,\infty}(\Omega)$ together with (3.5.8) and (3.5.9), we get the structural conditions

$$\langle \tilde{\boldsymbol{A}}(x,w,Dw), Dw\rangle \geq a_1|Dw|^2 - a_2 w^2, \qquad \tilde{B}(x,w) \leq 0,$$

valid when $w > 0$ and $x \in \Omega$; the details being essentially the same as in the derivation of (3.5.8).

Hence Theorem 3.2.2 implies $w \leq 0$ in Ω, that is $u \leq v$ in Ω. $\qquad \square$

This is essentially Theorem 10.7 (ii) of [43] with the important exceptions that $\boldsymbol{A}$ and B are allowed to be singular at $\boldsymbol{\xi} = \boldsymbol{0}$, and that the class $C^1(\Omega)$ is weakened to $W^{1,\infty}_{\text{loc}}(\Omega)$. Compare also Theorem 10.3 of [81].

3.6   Strongly degenerate operators

The condition of strict ellipticity in Theorem 3.5.1 can be avoided by adding suitable further hypotheses. This will allow us to cover the p-Laplace operator in the remaining case when $p > 2$, as well as general singular sets $\boldsymbol{Q}$. *We continue to assume conditions* (i) *and* (ii) *from the previous section, and furthermore*, except for Theorem 3.6.5, *the additional hypothesis*

(iii) *For all* $(x, z, \boldsymbol{\xi})$, $(x, z, \boldsymbol{\eta}) \in \Omega \times \mathbb{R} \times \mathbb{R}^n$ *we have*

$$\langle \boldsymbol{A}(x, z, \boldsymbol{\xi}) - \boldsymbol{A}(x, z, \boldsymbol{\eta}), \boldsymbol{\xi} - \boldsymbol{\eta} \rangle \geq 0.$$

An obvious case when (iii) occurs is the Euler–Lagrange equation for the variational integral

$$I[u] = \int_\Omega \mathscr{G}(x, u, Du)dx,$$

in which the integrand $\mathscr{G}(x, u, \boldsymbol{\xi})$ is convex in $\boldsymbol{\xi}$ but not strongly convex; that is, its gradient at some places is either too "*flat*" or has corners, e.g., the integrand $|Du|^p$ for $p \neq 2$.

Condition (iii) is automatic in the important case when $\boldsymbol{Q} = \{\mathbf{0}\}$ (or $\boldsymbol{Q} = \emptyset$) and $\boldsymbol{A} = \boldsymbol{A}(x, \boldsymbol{\xi})$ is elliptic in $\Omega \times \boldsymbol{P}$, as a consequence of Proposition 2.4.3.

The main comparison theorem for strongly degenerate elliptic inequalities is then the following

Theorem 3.6.1 (Comparison Principle). *Let* $\boldsymbol{P}$ *be a given open set in* $\mathbb{R}^n$. *Assume that* $\boldsymbol{A} = \boldsymbol{A}(x, \boldsymbol{\xi})$ *is independent of* z *and is elliptic in* $\Omega \times \boldsymbol{P}$. *Suppose also that* $B = B(x, z, \boldsymbol{\xi})$ *is locally Lipschitz continuous with respect to* $\boldsymbol{\xi}$ *in* $\Omega \times \mathbb{R} \times \mathbb{R}^n$ *and is non-increasing in the variable* z. *Let* u *and* v *be solutions of* (3.5.1) *and* (3.5.2) *of class* $W^{1,\infty}_{\mathrm{loc}}(\Omega)$, *such that*

$$\operatorname{essinf}_\Omega \{\operatorname{dist}(Du, \boldsymbol{Q}) + \operatorname{dist}(Dv, \boldsymbol{Q})\} > 0, \tag{3.6.1}$$

where $\boldsymbol{Q} = \mathbb{R}^n \setminus \boldsymbol{P}$. *If* $u \leq v + M$ *on* $\partial\Omega$, *where* M *is constant, then* $u \leq v + M$ *in* Ω.

Before giving the proof it is useful to establish the following

Lemma 3.6.2. *Let* $\boldsymbol{\xi}$, $\boldsymbol{\eta}$ *satisfy*

$$\boldsymbol{\xi}, \boldsymbol{\eta} \in \boldsymbol{B}_W, \qquad \operatorname{dist}(\boldsymbol{\xi}, \boldsymbol{Q}) + \operatorname{dist}(\boldsymbol{\eta}, \boldsymbol{Q}) \geq 4d$$

for some positive constants W *and* d, *with* $d \leq W$.

Let Γ be a compact subset of Ω. Then for all $x \in \Gamma \subset \Omega$ we have

$$\langle A(x, \xi) - A(x, \eta), \xi - \eta \rangle \geq a_1 |\xi - \eta|^2, \tag{3.6.2}$$

where

$$a_1 = \frac{d}{2W} \inf_{\Gamma \times \{P^d \cap B_W\}} \{\text{min eigenvalue of } [\partial_\xi A(x, \xi)]\}$$

and $P^d = \{\xi \in \mathbb{R}^n : \text{dist}(\xi, Q) \geq d\}$.

Proof. For $\xi \neq \eta$ we consider the line segment $[\xi, \eta]$, that is

$$\zeta(t) = (1 - t)\xi + t\eta, \qquad t \in [0, 1].$$

By hypothesis we may suppose without loss of generality that $\text{dist}(\eta, Q) \geq 2d$, so $\eta \in P^d$. There are two cases:

Case I. $[\xi, \eta] \not\subset P^d$; Case II. $[\xi, \eta] \subset P^d$.

In Case I, let $t_0 \in (0, 1)$ be such that $\zeta(t) \in P^d$ for all $t \in [t_0, 1)$ while $\text{dist}(\zeta(t_0), Q) = d$; see Figure 1. Then

$$I \equiv \langle A(x, \xi) - A(x, \eta), \xi - \eta \rangle$$
$$= \langle A(x, \xi) - A(x, \zeta_0), \xi - \eta \rangle + \langle A(x, \zeta_0) - A(x, \eta), \xi - \eta \rangle = I_1 + I_2,$$

where $\zeta_0 = \zeta(t_0)$. By (iii)

$$I_1 = \langle A(x, \xi) - A(x, \zeta_0), \xi - \zeta_0 \rangle \frac{|\xi - \eta|}{|\xi - \zeta_0|} \geq 0.$$

Moreover, since A is uniformly elliptic in $\Gamma \times \{P^d \cap B_W\}$, we have

$$I_2 = \langle A(x, \zeta_0) - A(x, \eta), \zeta_0 - \eta \rangle \frac{|\xi - \eta|}{|\zeta_0 - \eta|}$$
$$\geq a |\zeta_0 - \eta|^2 \frac{|\xi - \eta|}{|\zeta_0 - \eta|} = a |\xi - \eta|^2 \frac{|\zeta_0 - \eta|}{|\xi - \eta|},$$

where

$$a = \inf_{\Gamma \times \{P^d \cap B_W\}} \{\text{min eigenvalue of } [\partial_\xi A(x, \xi)]\}.$$

Finally, $|\zeta_0 - \eta| \geq d$ and $|\zeta_0 - \eta|/|\xi - \eta| \geq d/2W$, so that

$$I \geq I_2 \geq \frac{ad}{2W} |\xi - \eta|^2,$$

proving (3.6.2) for Case I.

Case II is obvious, with $I \geq a|\xi - \eta|^2 \ (\geq (ad/W)|\xi - \eta|^2)$. $\qquad \square$

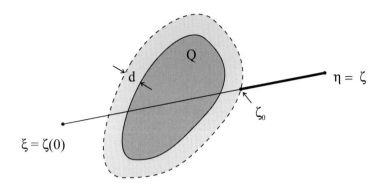

Figure 3.1: The set $\boldsymbol{P}^d$ is the complement of the shaded regions. Note that $[\boldsymbol{\zeta}_0, \boldsymbol{\eta}] \subset \boldsymbol{P}^d$ and $|\boldsymbol{\zeta}_0 - \boldsymbol{\eta}| \geq d$.

In the special case of the p-Laplace operator we can take $a_1 = d^{p-1}/2W$ when $p > 2$, while we have already shown that $a_1 = (p-1)W^{p-2}$ when $1 < p \leq 2$.

Proof of Theorem 3.6.1. With $M = 0$, and following Step 2 of the proof of Theorem 3.5.1 almost word-for-word, we see first that the function $w = u - v$ satisfies the inequality (3.5.7) with $w = 0$ on $\partial\Omega$. Also by (3.6.1) there is a number $d > 0$ such that

$$\text{essinf}_\Omega \left\{ \text{dist}(Du, \boldsymbol{Q}) + \text{dist}(Dv, \boldsymbol{Q}) \right\} \geq 4d.$$

Then since u, $v \in W^{1,\infty}_{\text{loc}}(\Omega)$ it follows from Lemma 3.6.2 that the operator $\tilde{A}$ in (3.5.7) satisfies the first structural condition of (3.2.1) along the solution w, that is with $\boldsymbol{\xi} = Dw$, and with $a_2 = 0$. Also as in Step 1 of Theorem 3.5.1 the function $\tilde{B}$ satisfies the second condition of (3.2.1) with $z = w$, $\boldsymbol{\xi} = Dw$, and with $b_2 = 0$.

The proof is now completed by applying Theorem 3.2.1. □

Corollary 3.6.3. *Assume that $B = B(x, z, \boldsymbol{\xi})$ is locally Lipschitz continuous with respect to $\boldsymbol{\xi}$ in $\Omega \times \mathbb{R} \times \mathbb{R}^n$ and is non-increasing in z. Let u and v be solutions of class $W^{1,\infty}_{\text{loc}}(\Omega)$ of the inequalities*

$$\Delta_p u + B(x, u, Du) \geq 0, \qquad \Delta_p v + B(x, v, Dv) \leq 0 \qquad \text{in } \Omega,$$

where $p > 1$. Suppose that

$$\text{essinf}_\Omega \{|Du| + |Dv|\} > 0.$$

If $u \leq v + M$ on $\partial\Omega$, where $M \geq 0$ is constant, then $u \leq v + M$ in Ω.

The next result is similar to Theorem 3.5.3 with the exception that A is not assumed to be uniformly elliptic and may depend on z.

Theorem 3.6.4 (Comparison Principle). *Let A be elliptic in $\Omega \times \mathbb{R} \times P$. Assume additionally that $B = B(x, z)$ does not depend on ξ and is non-increasing in the variable z. Let u and v be solutions of (3.5.1) and (3.5.2) of class $W^{1,\infty}(\Omega)$, with*

$$\operatorname{essinf}_\Omega \{\operatorname{dist}(Du, Q) + \operatorname{dist}(Dv, Q)\} > 0.$$

If $u \le v$ on $\partial\Omega$, then $u \le v$ in Ω.

Proof. The proof is a combination of the ideas of Theorems 3.5.3 and 3.6.1. $\qquad\square$

When the solutions u and v of the inequalities (3.5.1) and (3.5.2) are of class $C^1(\Omega)$, rather than $W^{1,\infty}_{\text{loc}}(\Omega)$, the hypotheses of Theorem 3.6.1 can be weakened, giving the second main result of the section.

Theorem 3.6.5 (Comparison Principle). *Let P be a given open set in $\mathbb{R}^n$. Assume that $A = A(x, \xi)$ is independent of z, obeys the conditions (i), (ii) stated in the previous section, and is elliptic in $\Omega \times P$. Suppose also that $B = B(x, z, \xi)$ is locally Lipschitz continuous with respect to ξ in $\Omega \times \mathbb{R}^+ \times \mathbb{R}^n$ and is non-increasing in the variable z.*

Let u, $v \in C^1(\Omega)$ be respectively solutions of the inequalities (3.5.1) and (3.5.2) in the subsets

$$\Omega_u = \{x \in \Omega : Du(x) \in P\}, \qquad \Omega_v = \{x \in \Omega : Dv(x) \in P\}.$$

Assume finally that $\Omega_u \cup \Omega_v = \Omega$ and that $u \le v + M$ on $\partial\Omega$, M constant. Then $u \le v + M$ in Ω.

Proof. It is enough to consider the case $M = 0$. Thus suppose for contradiction that $u > v$ at some point in Ω. Let

$$V = \max_\Omega (u - v) > 0$$

be the supremum of $u - v$ in Ω, this being attained at an interior point y since $u - v \le 0$ on $\partial\Omega$. Of course $D(u - v) = 0$ at y, so from the condition $\Omega_u \cup \Omega_v = \Omega$ it follows that $Du(y) = Dv(y) \in P$.

Also let

$$\Sigma = \{x \in \Omega : \ell < u - v \le V\}, \qquad \ell \in (0, V),$$

be a neighborhood of the critical point y. Since Du and Dv can be made arbitrarily near $Du(y)$ in Σ by fixing ℓ sufficiently near V, we obtain $Du \in P$, $Dv \in P$ in Σ. In particular $\Sigma \subset \Omega_u \cap \Omega_v$, so u and v are solutions of (3.5.1) and (3.5.2) in the set Σ. In turn the comparison Theorem 3.5.1 can be applied to the solutions u and v in Σ. In fact Du, Dv can be supposed to lie in a compact subset N of P, with the consequence that A is strictly elliptic in $\Sigma \times N$ and the regular case of Theorem 3.5.1 is applicable.

Since $u = v + \ell$ on $\partial\Sigma$, it follows that $u \leq v + \ell$ in Σ. That is, $u - v \leq \ell < V$ in Σ, which contradicts the fact that $u - v = V$ at y. $\square$

3.7 Maximum principles for non-homogeneous elliptic inequalities

Consider the differential operator

$$L[u] = \operatorname{div}\boldsymbol{A}(x, u, Du) + B(x, u, Du),$$

where

$$\boldsymbol{A} : \Omega \times \mathbb{R} \times \mathbb{R}^n \to \mathbb{R}^n, \qquad B : \Omega \times \mathbb{R} \times \mathbb{R}^n \to \mathbb{R}$$

and $\boldsymbol{A}$ satisfies the hypotheses (i)–(iii) of Sections 3.5 and 3.6. Additionally we assume

(iv) $\partial_z \boldsymbol{A}(x, z, \boldsymbol{\xi})$ is locally Lipschitz continuous with respect to $\boldsymbol{\xi}$ in $\Omega \times \mathbb{R} \times \mathbb{R}^n$.

Theorem 3.6.1 has as one of its main consequence the following maximum principle for non-homogenous elliptic inequalities. It is interesting that for this result the function $B(x, z, \boldsymbol{\xi})$ *need not* be monotone in the variable z.

Theorem 3.7.1 (Maximum Principle). *Assume that $\boldsymbol{A} = \boldsymbol{A}(x, z, \boldsymbol{\xi})$ is elliptic in $\Omega \times \mathbb{R}^+ \times P$ and that $B = B(x, z, \boldsymbol{\xi})$ is locally Lipschitz continuous with respect to $\boldsymbol{\xi}$ in $\Omega \times \mathbb{R}^+ \times \mathbb{R}^n$.*

Define $\mathscr{L}[z, v] : \mathbb{R}^+ \times C^1(\Omega) \to \mathbb{R}$ pointwise by

$$\mathscr{L}[z, v](x) = \operatorname{div}\boldsymbol{A}(x, z, Dv) + \mathscr{B}(x, z, Dv), \qquad (3.7.1)$$

where

$$\mathscr{B}(x, z, \boldsymbol{\xi}) = \langle \partial_z \boldsymbol{A}(x, z, \boldsymbol{\xi}), \boldsymbol{\xi} \rangle + B(x, z, \boldsymbol{\xi})$$

for all $x \in \Omega$, $z \in \mathbb{R}^+$ and $\boldsymbol{\xi} \in \mathbb{R}^n$.

Let $v = v(x) \in C^1(\Omega)$ be a non-negative comparison function *for the operator L, in the sense that* $v(x) \geq 0$ *and* $Dv(x) \in \boldsymbol{P}$ *for* $x \in \Omega$; *and* $\mathscr{L}[z, v] \leq 0$ *for all* $z > 0$. *If* $u \in W^{1,\infty}_{loc}(\Omega)$ *is a solution of the inequality* $L[u] \geq 0$ *in* Ω *and* $u \leq v$ *on* $\partial\Omega$, *then* $u \leq v$ *in* Ω.

Proof. Define

$$\tilde{\mathscr{L}}[v] \equiv \operatorname{div}\boldsymbol{A}(x, u(x), Dv) + \langle \partial_z \boldsymbol{A}(x, u(x), Dv), Dv - Du \rangle + B(x, u(x), Dv).$$

By direct calculation one gets

$$\operatorname{div}\boldsymbol{A}(x, u(x), Dv) = \operatorname{div}\boldsymbol{A}(x, z, Dv) + \langle \partial_z \boldsymbol{A}(x, z, Dv), Du \rangle$$

evaluated at $z = u(x)$ *in* Ω. Hence in Ω

$$\begin{aligned}
\tilde{\mathscr{L}}[v] &= \left[\operatorname{div}\boldsymbol{A}(x, z, Dv) + \langle \partial_z \boldsymbol{A}(x, z, Dv), Dv \rangle + B(x, z, Dv) \right]_{z=u(x)} \\
&= \mathscr{L}[z, v]_{z=u(x)}.
\end{aligned}$$

By hypothesis, then, $\tilde{\mathscr{L}}[v] \leq 0$ whenever $u(x) > 0$. On the other hand, clearly $\tilde{\mathscr{L}}[u] \geq 0$ in Ω. From its definition we see that $\tilde{\mathscr{L}}[v]$ can be written in the form

$$\tilde{\mathscr{L}}[v] = \operatorname{div}\tilde{\boldsymbol{A}}(x, Dv) + \tilde{B}(x, Dv), \tag{3.7.2}$$

where

$$\begin{aligned}
\tilde{\boldsymbol{A}}(x, \boldsymbol{\xi}) &= \boldsymbol{A}(x, u(x), \boldsymbol{\xi}), \\
\tilde{B}(x, \boldsymbol{\xi}) &= \langle \partial_z \boldsymbol{A}(x, u(x), \boldsymbol{\xi}), \boldsymbol{\xi} - Du \rangle + B(x, u(x), \boldsymbol{\xi}).
\end{aligned}$$

Of course both $\tilde{\boldsymbol{A}}$ and $\tilde{B}$ are independent of z. Therefore in view of (i)–(iv) the functions $\tilde{\boldsymbol{A}}$ and $\tilde{B}$ satisfy conditions (i)–(iii), while $\tilde{B}$ is locally Lipschitz continuous with respect to $\boldsymbol{\xi}$ in $\Omega \times \mathbb{R}^+ \times \mathbb{R}^n$.

Finally $\tilde{\boldsymbol{A}}(x, \boldsymbol{\xi})$ is elliptic when $x \in \Omega$, $u(x) > 0$ and $\boldsymbol{\xi} \in \boldsymbol{P}$, since $\boldsymbol{A}(x, z, \boldsymbol{\xi})$ is elliptic in $\Omega \times \mathbb{R}^+ \times \boldsymbol{P}$.

Let $\Omega' = \{x \in \Omega : u(x) > 0\}$. It is easy to see that $u \leq v$ on $\partial\Omega'$. We can apply Theorem 3.6.1 to any component $\mathscr{C}$ of Ω', with u and v satisfying $\tilde{\mathscr{L}}[u] \geq 0$, $\tilde{\mathscr{L}}[v] \leq 0$, $Dv \in \boldsymbol{P}$ in $\mathscr{C}$ (so $\operatorname{dist}(Dv, \boldsymbol{Q}) > 0$). Hence $u \leq v$ in $\mathscr{C}$ and in turn $u \leq v$ in Ω'. This finally gives $u \leq v$ in Ω, completing the proof. ☐

Theorem 3.7.1 is somewhat abstract, in that it depends on the *existence* of the comparison function v. As in Theorem 2.3.2, when $\boldsymbol{A}$ and B are more specialized we can avoid this difficulty. In particular, consider the case where $Q \subset B_\varrho$ for some $\varrho \geq 0$ (the possibility $P = \mathbb{R}^n$ is included when $\varrho = 0$). Assume that

$$\begin{cases} A(x, z, \boldsymbol{\xi}) \text{ is elliptic,} \\ B(x, z, \boldsymbol{\xi}) + A^*(x, z, \boldsymbol{\xi}) \leq \alpha|\boldsymbol{\xi}| \, E(x, z, \boldsymbol{\xi}) + \gamma \end{cases} \qquad (3.7.3)$$

in $\Omega \times \mathbb{R}^+ \times P$, where α and γ are non-negative constants, and

$$A^*(x, z, \boldsymbol{\xi}) = \text{Trace}\,[\partial_x A(x, z, \boldsymbol{\xi})] + \langle \partial_z A(x, z, \boldsymbol{\xi}), \boldsymbol{\xi} \rangle,$$
$$E(x, z, \boldsymbol{\xi}) = \partial_{\xi_i} A_j(x, z, \boldsymbol{\xi}) \frac{\xi_i \xi_j}{|\boldsymbol{\xi}|^2}. \qquad (3.7.4)$$

Note that $A^* = 0$ in the important case when $A = A(\boldsymbol{\xi})$.

Theorem 3.7.2 (Maximum Principle). *Let A and B satisfy* (3.7.3), *and suppose that*

$$|\boldsymbol{\xi}| \, E(x, z, \boldsymbol{\xi}) \geq \Psi(|\boldsymbol{\xi}|) \qquad \text{in } \Omega \times \mathbb{R}^+ \times P, \qquad P = \mathbb{R}^n \setminus Q, \quad (3.7.5)$$

where $\Psi = \Psi(t)$ is a strictly increasing function on (ϱ, ∞), $\varrho \geq 0$.

Let u be a solution of class $W_{\text{loc}}^{1,\infty}(\Omega)$ of the boundary value problem

$$\begin{array}{ll} \text{div}\, A(x, u, Du) + B(x, u, Du) \geq 0 & \text{in } \Omega, \\ u \leq 0 & \text{on } \partial\Omega, \end{array} \qquad (3.7.6)$$

where $\Omega \subset \{x \in \mathbb{R}^n \,:\, 0 < x_1 < R\}$. Then there holds

$$u(x) \leq R \max\{\rho, C\}(e^k - 1), \qquad (3.7.7)$$

where[3]

$$C = \Psi^{-1}(R\gamma), \quad k = 1 + \alpha R, \text{ when } \lim_{t\to\infty} \Psi(t) > 2\gamma R,$$
$$C = \Psi^{-1}(\ell), \qquad k = 1 + (\alpha + \gamma/\ell)R, \qquad (3.7.8)$$
$$\text{when } \lim_{t\to\infty} \Psi(t) = 2\ell \leq 2\gamma R.$$

Theorem 3.7.2 has almost exactly the formulation of the earlier Theorem 2.3.2. For completeness the full proof is given here, even though it is essentially the same as for the earlier result.

---

[3]If $\Psi(\varrho) = \lim_{t\to\varrho+} \Psi(t) = \ell' > 0$, then we define $\Psi^{-1}(s) = \varrho$ when $s \leq \ell'$. Note that the case $\lim_{t\to\infty} \Psi(t) < \infty$ is possible. That is, take $A(\boldsymbol{\xi}) = 2\ell \, \log(|\boldsymbol{\xi}| + 1) \dfrac{\boldsymbol{\xi}}{|\boldsymbol{\xi}|}$ and use the computation of footnote 3 of Section 2.2.

Proof. It is enough to construct a comparison function $v = v(x)$ such that $v(x) > 0$ in Ω and $\mathscr{L}[z, v] \leq 0$ for all $z > 0$. Accordingly, we choose

$$v(x) = K(e^{mR} - e^{mx_1}), \qquad x \in \Omega,$$

where $m = k/R$, $K > R\max\{\varrho, C\}$. Then $\partial_{x_1} v(x) = -Kme^{mx_1}$ so $|Dv| \geq mK \geq (1 + \alpha R)\varrho$. Also

$$\partial_{x_1}^2 v(x) = -Km^2 e^{mx_1} = -m|Dv|.$$

In view of (3.7.1) and (3.7.3), a direct calculation then shows that $\mathscr{L}[z, v] \leq 0$ in Ω provided

$$m|Dv|\partial_{\xi_1} A_1(x, z, Dv) \geq \alpha|Dv|E(x, z, Dv) + \gamma. \qquad (3.7.9)$$

But $E(x, z, Dv) = \partial_{\xi_1} A_1(x, z, Dv)$, so (3.7.9) becomes

$$m|Dv|E(x, z, Dv) \geq \alpha|Dv|E(x, z, Dv) + \gamma. \qquad (3.7.10)$$

Obviously (3.7.10) is satisfied if $(m - \alpha)|Dv|\, E(x, z, Dv) \geq \gamma$ for all $z > 0$. At the same time

$$|Dv|E(x, z, Dv) \geq \Psi(|Dv|) \geq \Psi(mK) \geq \Psi(C) \geq \min\{\gamma R, \ell\},$$

since $mK > (k/R)R\max\{\varrho, C\} \geq C$. Therefore (3.7.10) holds when k and C are given as in (3.7.8), and in turn we get $\mathscr{L}[z, v] \leq 0$ in Ω, as required. We now apply Theorem 3.7.1, giving

$$u(x) \leq v(x) \leq K(e^k - 1) \qquad \text{in } \Omega.$$

Letting $K \to R\max\{\varrho, C\}$ completes the proof. $\qquad\qquad\square$

The remarks after Theorem 2.3.2 apply equally to the previous result. When B is homogeneous the global condition (3.7.3) need be assumed only for $|\xi|$ small. We state this result as

Theorem 3.7.3. *Assume $\boldsymbol{P} = \mathbb{R}^n$ or $\boldsymbol{P} = \mathbb{R}^n \setminus \{0\}$. Let the hypotheses of Theorem 3.7.2 hold, with the exceptions that $\gamma = 0$, and (3.7.3) and (3.7.5) are assumed to be valid only in $\Omega \times \mathbb{R}^+ \times \boldsymbol{R}_1$. Let u be a solution of class $W_{\text{loc}}^{1,\infty}(\Omega)$ of the boundary value problem (3.7.6) where Ω is a bounded domain in $\mathbb{R}^n$. Then $u \leq 0$ in Ω.*

In the generality of the present hypotheses, this seems to be a new result.

Proof. Since $\gamma = 0$ only the first case of (3.7.8) applies and so $C = \Psi^{-1}(0) = \varrho = 0$. In this case the constant $K > 0$ in the proof of Theorem 3.7.2 can be chosen arbitrarily small, and in particular so that $|Dv(x)| \leq K m e^{mR} \leq 1$ in Ω. The rest of the proof of Theorem 3.7.2 then applies without change, giving $u \leq 0$, independent of R. Since Ω is bounded we get $u \leq 0$ in Ω. □

Theorem 3.7.3 is false if one weakens condition (3.7.3); see the comment after Theorem 2.3.3 and example (2.3.9).

Theorem 3.7.2 has a further direct application.

Theorem 3.7.4 (Maximum Principle). *Let $A \in C^1(\mathbb{R}^+)$, $A(s) > 0$ and $\Lambda(s) = s[A(s) + sA'(s)] > 0$ for $s > 0$. Assume that Λ is strictly increasing, $\Lambda(0) = 0$, and, for simplicity, also that $\Lambda(s) \to \infty$ as $s \to \infty$. Suppose finally that*

$$B(x, z, \boldsymbol{\xi}) \leq \alpha \Lambda(|\boldsymbol{\xi}|) + \gamma \qquad in \ \Omega \times \mathbb{R}^+ \times (\mathbb{R}^n \setminus \{\mathbf{0}\}),$$

where α and γ are non-negative constants.

Let u be a solution of class $W_{loc}^{1,\infty}(\Omega)$ of the boundary value problem

$$\begin{aligned} \operatorname{div}\{A(|Du|)Du\} + B(x, u, Du) &\geq 0 & in \ \Omega, \\ u &\leq 0 & on \ \partial\Omega, \end{aligned} \tag{3.7.11}$$

where $\Omega \subset \{x \in \mathbb{R}^n \ 0 < x_1 < R\}$. Then there holds

$$u(x) \leq R\Lambda^{-1}(R\gamma)\left[e^{1+\alpha R} - 1\right].$$

Furthermore, when $\gamma = 0$ then $u \leq 0$ in Ω, where Ω can be any bounded domain in $\mathbb{R}^n$.

Proof. In the present case $\boldsymbol{Q} = \{\mathbf{0}\}$, and

$$\boldsymbol{A}(\boldsymbol{\xi}) = A(|\boldsymbol{\xi}|)\boldsymbol{\xi}, \qquad \partial_{\boldsymbol{\xi}}\boldsymbol{A}(\boldsymbol{\xi}) = A(|\boldsymbol{\xi}|)\mathbb{I}_n + A'(|\boldsymbol{\xi}|)\frac{\boldsymbol{\xi} \otimes \boldsymbol{\xi}}{|\boldsymbol{\xi}|}, \qquad \boldsymbol{A}^*(\boldsymbol{\xi}) = \mathbf{0},$$

with $\boldsymbol{A}^*$ defined in (3.7.4). The eigenvalues of the Jacobian matrix are $A(s)$ and $A(s) + sA'(s)$, with $s = |\boldsymbol{\xi}|$. Therefore by hypothesis the equation (3.7.11) is elliptic for $\boldsymbol{\xi} \neq \mathbf{0}$. It is easy to see moreover that $E(\boldsymbol{\xi}) = A(s) + sA'(s)$, and in turn $|\boldsymbol{\xi}| E(\boldsymbol{\xi}) = \Lambda(s)$. The conclusion is now immediate from Theorem 3.7.2, with $\boldsymbol{Q} = \{\mathbf{0}\}$ and $\Psi(s) = \Lambda(s)$.

The final statement of the theorem is obvious from the previous proof. □

Remarks

1. When $A(s) = s^{p-2}$, $p > 1$, we get the important subcase of the p-Laplace operator, for which $E(s) = (p-1)s^{p-2}$, $\Lambda(s) = (p-1)s^{p-1}$ and $R\Lambda^{-1}(Rs) = [s/(p-1)]^{1/(p-1)} R^{p'}$. See also the comments after Theorem 2.3.2.

2. The possibility that $Q \supsetneq \{0\}$, say $Q = B_\varrho$, $\varrho > 0$, in Theorem 3.7.2 can be illustrated by the example

$$A(\boldsymbol{\xi}) = \begin{cases} 0, & \text{if } |\boldsymbol{\xi}| \leq 1, \\ |\boldsymbol{\xi}|^{p-2}\boldsymbol{\xi} - |\boldsymbol{\xi}|^{-1}\boldsymbol{\xi}, & \text{if } |\boldsymbol{\xi}| \geq 1, \end{cases} \tag{3.7.12}$$

with $p > 1$. Clearly $\boldsymbol{A}$ satisfies the basic conditions (i), (ii) and (iv), together with the hypothesis (3.7.3) of Theorem 3.7.2 with $\varrho = 1$. In (3.7.4) we have $\boldsymbol{A}^* = \boldsymbol{0}$ and

$$E(x, z, \boldsymbol{\xi}) = E(\boldsymbol{\xi}) = (p - 1)|\boldsymbol{\xi}|^{p-2}, \qquad \text{if } |\boldsymbol{\xi}| \geq 1.$$

Thus in turn

$$\Psi(s) = (p - 1)s^{p-1} \qquad \text{if } s \geq 1,$$

which is strictly increasing in $[1, \infty)$ and tends to ∞ as $s \to \infty$. The principal condition (3.7.3) then becomes

$$B(x, z, \boldsymbol{\xi}) \leq \alpha|\boldsymbol{\xi}|^{p-1} + \gamma, \qquad \text{if } |\boldsymbol{\xi}| \geq 1,$$

with no restriction assigned when $|\boldsymbol{\xi}| \leq 1$, namely in $\boldsymbol{B}_1$. Of course for the applicability of Theorem 3.7.2 the remaining assumption (iii) must be required on B.

The conclusion is

$$u(x) \leq \max\{1, \gamma^{1/(p-1)}\} \cdot [e^{1+\alpha R} - 1].$$

Obviously results of this kind do not follow from the theory in [43].

Theorem 3.7.5. *Let the hypotheses of Theorem 3.7.2 be satisfied, with the exception that (3.7.3) is replaced by the condition that*

$$B(x, z, \boldsymbol{\xi}) + \boldsymbol{A}^*(x, z, \boldsymbol{\xi}) \leq (\alpha|\boldsymbol{\xi}| + \beta|\boldsymbol{\xi}|^q)E(x, z, \boldsymbol{\xi}) + \gamma, \qquad 0 < q < 1,$$

in $\Omega \times \mathbb{R}^+ \times \boldsymbol{P}$, where α, β, γ are non-negative constants.

Then (3.7.7) holds with the previous constant C replaced by $C + \beta^{1/(1-q)}$ and the previous constant k replaced by $k + 1$.

The proof is essentially the same as before. The additional term $\gamma|\boldsymbol{\xi}|^q$ (in the case $q = 0$) was first introduced by Gilbarg and Trudinger ([43], Theorem 10.3).

3.8 Uniqueness of the singular Dirichlet problem

The structure built up in the earlier parts of this chapter allows one to present a number of uniqueness theorems for distribution solutions of the Dirichlet problem

$$\operatorname{div} \boldsymbol{A}(x, u, Du) + B(x, u, Du) = 0 \qquad \text{in } \Omega,$$
$$u = u_0 \qquad \text{on } \partial\Omega, \tag{3.8.1}$$

where $u_0 \in C(\partial\Omega)$ and Ω is a bounded domain of $\mathbb{R}^n$. We assume that $\boldsymbol{A}$ and B satisfy the hypotheses (i)–(iv) of Sections 3.5–3.7.

Theorem 3.8.1. *Suppose that $\boldsymbol{A} = \boldsymbol{A}(x, \boldsymbol{\xi})$ is independent of z and strictly elliptic in $\Omega \times \boldsymbol{R}_1$.*

Assume additionally that $B = B(x, z, \boldsymbol{\xi})$ is locally Lipschitz continuous with respect to $\boldsymbol{\xi}$ in $\Omega \times \mathbb{R}^+ \times \boldsymbol{P}$ and is non-increasing in the variable z.

Then problem (3.8.1) can have at most one solution of class $W^{1,\infty}_{\text{loc}}(\Omega)$.

This is an immediate consequence of Theorem 3.5.1.

Theorem 3.8.2. *Assume that $\boldsymbol{A} = \boldsymbol{A}(x, \boldsymbol{\xi})$ is independent of z and is elliptic in $\Omega \times \boldsymbol{P}$. Suppose also that B is non-increasing in z. Let u and v be solutions of class $W^{1,\infty}_{\text{loc}}(\Omega)$ of (3.8.1), with*

$$\operatorname{essinf}_\Omega \{\operatorname{dist}(Dv, \boldsymbol{Q}) + \operatorname{dist}(Du, \boldsymbol{Q})\} > 0.$$

Then $u = v$ in Ω.

This is a corollary of Theorem 3.6.1

In the same way the Comparison Theorems 3.5.3 and 3.6.4 allow corresponding uniqueness results, whose statements can be left to the reader. The special case of the p-Laplace operator is of particular interest.

Corollary 3.8.3. *Let $B = B(x, z, \boldsymbol{\xi})$ be non-increasing in the variable z. Let u and v be solutions of class $W^{1,\infty}_{\text{loc}}(\Omega)$ of the Dirichlet problem*

$$\Delta_p u + B(x, u, Du) = 0 \qquad \text{in } \Omega,$$
$$u = u_0 \qquad \text{on } \partial\Omega, \tag{3.8.2}$$

where $u_0 \in C(\partial\Omega)$.

Then $u = v$ if $1 < p \leq 2$ and B is regular. The same conclusion holds when $p > 2$ (without the condition that B be regular), provided that either $\operatorname{essinf}_\Omega |Du| > 0$ or $\operatorname{essinf}_\Omega |Dv| > 0$.

This is an obvious consequence of Corollaries 3.5.2 and 3.6.3.

Remarks

1. The second part of Corollary 3.8.3 fails when $\mathrm{essinf}_\Omega\{|Du|+|Dv|\}=0$. Indeed the problem

$$\Delta_4 u + |Du|^2 = 0 \quad \text{in } B_R \subset \mathbb{R}^2, \qquad u = 0 \quad \text{on } \partial B_R,$$

admits the two solutions $u(x) = 0$ and $v(x) = \frac{1}{8}(R^2 - |x|^2)$ in B_R. Here $|Du| + |Dv| = 0$ at 0, and in turn $\mathrm{essinf}_\Omega\{|Du| + |Dv|\} = 0$.

2. As an application of the second part of Corollary 3.8.3, consider the problem (3.8.2), with $B(x, z, \boldsymbol{\xi}) = |\boldsymbol{\xi}|^2 - 1$ and $u_0(x) = x_1$, $x = (x_1, \ldots, x_n)$. This admits only the single solution $u(x) = x_1$ whatever the bounded domain Ω may be, since $|Du| = 1$ in $\mathbb{R}^n$.

When the boundary data takes the canonical form $u = 0$ on $\partial\Omega$, then the condition in Theorem 3.8.1 that $\boldsymbol{A}$ be strictly elliptic can be dropped. The result is as follows.

Theorem 3.8.4. *Let* $\boldsymbol{A}(x, z, \boldsymbol{\xi})$ *be elliptic in* $\Omega \times \mathbb{R} \times \boldsymbol{P}$, *where* $\boldsymbol{Q} = \emptyset$ *or* $\{\boldsymbol{0}\}$. *Assume that*

$$[\mathrm{sign}\, z] \cdot B(x, z, \boldsymbol{\xi}) \le \alpha \Psi(|\boldsymbol{\xi}|), \tag{3.8.3}$$

with $|\boldsymbol{\xi}|E(x, z, \boldsymbol{\xi}) \ge \Psi(|\boldsymbol{\xi}|)$, *where* Ψ *is strictly increasing in* $\mathbb{R}$, $\Psi(0) = 0$, *and* E *is given by* (3.7.4).
If $u \in W^{1,\infty}_{\mathrm{loc}}(\Omega)$ *is a solution of the Dirichlet problem*

$$\begin{array}{ll} \mathrm{div}\, \boldsymbol{A}(x, u, Du) + B(x, u, Du) = 0 & \text{in} \quad \Omega, \\ u = 0 & \text{on } \partial\Omega, \end{array} \tag{3.8.4}$$

then $u \equiv 0$.

Proof. This follows immediately from Theorem 3.7.3, when we observe that the function $v(x) = -u(x)$ also satisfies an equation of the form (3.8.4), with the corresponding inequality (3.8.3) equally valid; that is, the only possible solution of (3.8.4) is $u \equiv 0$. $\qquad\square$

3.9   Appendix: Sobolev's inequality

Here we review various results which are needed in the earlier parts of the chapter. We begin with the standard Sobolev inequality.

Theorem 3.9.1 (Theorem 7.10 of [43]). *Let $1 \leq p < n$. Then there exists a constant $S(p,n)$ such that for every function $u \in W_0^{1,p}(\Omega)$, $\Omega \subset \mathbb{R}^n$, such that*

$$\|u\|_{p^*,\Omega} \leq S(p,n)\|Du\|_p, \qquad p^* = \frac{np}{n-p}.$$

An explicit bound for $S(p,n)$ is given in [43], that is $S(p,n) \leq (n-1)p/\sqrt{n}(n-p)$. This is less than 1 for p suitably near 1. The case $p = 1$ is particularly simple: $S(1,n) = n^{-1}\omega^{-1/n}$, see [107] and also [37], where the result is indicated rather obscurely.

The Sobolev inequality has another useful formulation.

Theorem 3.9.2. *Let n, $s \geq 1$, and*

$$p \geq \max\{1, ns/(n+s))\} \stackrel{\text{def}}{=} s_*.$$

Then

$$\|u\|_s \leq S(s_*,n)\|Du\|_p|\Omega|^{1/n-1/p+1/s}.$$

Proof. Suppose first that $s \geq n/(n-1)$. Then $1 \leq s_* < n$, $(s_*)^* = s$. Hence by the main Sobolev theorem

$$\|u\|_s \leq S(s_*,n)\|Du\|_{s_*} \leq S(s_*,n)|\Omega|^{1/s_*-1/p}\|Du\|_p$$

since $p \geq s_*$.

If $1 \leq s < n/(n-1)$, then $s_* = 1$, $(s_*)^* = n/(n-1)$ and so

$$\|u\|_s \leq \|u\|_{n/(n-1)}|\Omega|^{1/n-1+1/s} \leq S(1,n)\|Du\|_1|\Omega|^{1/n-1+1/s}$$
$$\leq S(1,n)\|Du\|_p|\Omega|^{1/n-1/p+1/s};$$

here the case $n = 1$ applies equally, since $S(1,1)$ is finite. $\qquad\square$

Note that if $s \to \infty$, then $s_* \to n$ and $S(s_*,n) \to \infty$.

Theorem 3.9.3 (Morrey inequality). *Let $p > n$ and $|\Omega| = 1$. There exists a constant $Q_\infty = Q_\infty(p,n)$ such that any function $u \in W_0^{1,p}(\Omega)$ has a continuous representative (still called u) such that*

$$\sup_\Omega |u| \leq Q_\infty \|Du\|_p.$$

A proof of Theorem 3.9.3 is given later, see the comments after Theorem 7.5.7. Finally, we have the simplest case of the Poincaré inequality.

Theorem 3.9.4 (Poincaré inequality). *Let $p \geq 1$. Then there exists a constant $Q = Q(n) = \omega_n^{-1/n}$ such that every function $u \in W_0^{1,p}(\Omega)$ obeys*

$$\|u\|_p \leq Q|\Omega|^{1/n}\|Du\|_p.$$

Theorem 3.9.4 is not best possible if Q is allowed to depend also on p. For example, if $p = 1$ we can use $Q = S(1, n) = n^{-1}\omega^{-1/n}$, as follows from Theorem 3.9.2 with $p = q = s = s_* = 1$.

Notes

The early results of this chapter, Theorem 3.2.1 and Theorem 3.2.2, are special cases of the later Theorems 6.1.3 and 6.1.4, but along with their proofs are of interest in themselves. The importance of thin sets theorems such as Theorem 3.3.1 seems to have been first pointed out by Berestycki and Nirenberg [11]. Thin set theorems, however, already appear in the work of Gilbarg and Trudinger [43], Theorem 10.10. It is worth adding that by using the differencing technique of Section 2.5 one can obtain thin set comparison theorems without monotonicity conditions on the function B.

Theorems 3.5.1 and 3.5.3 generalize the corresponding Theorem 10.7 (i) and (ii) of [43], in that we treat solutions in $W_{\text{loc}}^{1,\infty}(\Omega)$ rather than $C^1(\overline{\Omega})$, and also allow the operator A and the nonlinearity B to be singular (degenerate). Theorems 3.5.1 and 3.6.1 appear in weaker forms as Theorems 10.3 and 10.1 of [81]. Theorems 3.6.1, 3.6.4, 3.7.1–3.7.4 are new; it is interesting that they are the direct analogues of Theorems 2.2.3–2.3.3 for non-divergence operators.

Problems

3.1 Supply full details for the proof of the key Lemma 3.1.2. Discuss the importance of the p-regularity condition.

3.2 Show that the p-Laplace operator $A(\xi) = |\xi|^{p-2}\xi$, $p > 1$, is automatically p-regular.

3.3 Justify inequality (3.2.4).

3.4 Carry out the details for the proofs of Theorems 3.2.3 and 3.2.4.

3.5 Ditto for Corollary 3.3.2.

3.6 Ditto for Theorem 3.6.1.

3.7 Carry out the calculations required to prove the relations (3.7.9) and (3.7.10), and show that (3.7.10) holds when k and C are given by (3.7.8).

3.8 Check that the operator (3.7.12) satisfies the given conditions (i), (ii) and (iv), as well as condition (3.7.3), with $\varrho = 1$.

3.9 Consider the quasilinear equation [76, (11) on page 153]

$$[\mu(|Du|) - (\partial_y u)^2]\partial_{x^2}^2 u + 2\partial_x u \, \partial_y u \, \partial_{xy}^2 u + [\mu(|Du|) - (\partial_x u)^2]\partial_{y^2}^2 u = 0,$$

which arises in the study of the flow of compressible fluids, and carry out the details required for Protter and Weinberger's discussion of subsonic flow on pages 153–155 of [76]. (Note that the bold face statement on page 155 applies only to subsonic flow.)

Chapter 4

Boundary Value Problems for Nonlinear Ordinary Differential Equations

4.1 Preliminary lemmas

Here we begin the study of the strong maximum principle and the compact support principle for divergence structure inequalities, especially of the canonical form

$$\operatorname{div}\{A(|Du|)Du)\} - f(u) \leq 0, \qquad u \geq 0. \tag{4.1.1}$$

In general, the results described cannot be obtained from the nonlinear theorems of the previous chapters, since equation (4.1.1) has specialized properties which are crucially used.

We assume throughout, unless otherwise mentioned, that the functions A and f satisfy conditions $(A1)$, $(A2)$, $(F1)$, $(F2)$ in the introduction. Here $\Phi(s) = sA(s)$, $s > 0$, and $H = H(s)$ is the Legendre transform defined in $(1.1.4)$. For convenience in what follows it is useful to extend the definition of the principal operator Φ to all real values of s by setting $\Phi(s) = -\Phi(-s)$ when $s < 0$, unless otherwise explicitly specified. Without loss of generality, since we deal with non-negative solutions, one may suppose that

$$f(z) = 0 \quad \text{for} \quad z \leq 0.$$

We start with a series of preliminary results, drawn from [81].

Lemma 4.1.1.

(i) *For any constant $\sigma \in [0, 1]$ there holds*

$$F(\sigma z) \leq \sigma F(z), \quad z \in [0, \delta); \qquad F(z) = \int_0^z f(v)dv.$$

(ii) *Let $w = w(t)$ be of class $C^1(0, T)$, and write $' = d/dt$. If the composition $\Phi \circ w'$ is of class $C^1(0, T)$, then $H \circ w'$ is of class $C^1(0, T)$, and in this case*

$$[H(w'(t))]' = w'(t)[\Phi(w'(t))]' \qquad in \ (0, T). \tag{4.1.2}$$

Conversely, if $H \circ w'$ is of class $C^1(0, T)$ and $w' > 0$, then $\Phi \circ w'$ is of class $C^1(0, T)$ and (4.1.2) continues to be satisfied.

To obtain (i), observe that $\sigma f(\sigma z) \leq \sigma f(z)$ for $z \in [0, \delta)$, since f is non-decreasing by (F2). Integrating this relation from 0 to z yields the result.

The first statement of (ii) is an immediate consequence of the representation

$$H(s) = \int_0^{\Phi(s)} \Phi^{-1}(w)dw, \qquad s \geq 0, \tag{4.1.3}$$

this being a consequence of the Stieltjes formula $H(s) = \int_0^s \sigma \, d\Phi(\sigma)$. The second part is also a consequence of (4.1.3) together with a small lemma:

Let I be any interval of $\mathbb{R}$ and let

$$B(t) = \int_{a(t_0)}^{a(t)} b(s)ds, \qquad t, t_0 \in I.$$

Suppose $a, b \in C(I)$, $B \in C^1(I)$ and $b > 0$. Then $a \in C^1(I)$ and $a' = B'/(b \circ a)$.

This is easily demonstrated by using difference coefficients and the integral mean value theorem to get $\Delta B/\Delta t = b(a + \theta \Delta a)\Delta a/\Delta t$, $0 \leq \theta \leq 1$. The lemma then follows by dividing by $b(a + \theta \Delta a)$ and letting $\Delta t \to 0$.

Lemma 4.1.2. *Suppose $f(z) > 0$ for $z > 0$. Let $\sigma > 0$. If (1.1.7) is satisfied, then*

$$\int_{0+} \frac{ds}{H^{-1}(\sigma F(s))} < \infty,$$

while if (1.1.5) *holds, then*

$$\int_{0+} \frac{ds}{H^{-1}(\sigma F(s))} = \infty.$$

Proof. To show the first part of the lemma it is obviously enough to consider values $\sigma < 1$. In this case, by Lemma 4.1.1 (i), and with δ chosen such that $F(\delta) < H(\infty)$,

$$\int_\varepsilon^\delta \frac{ds}{H^{-1}(\sigma F(s))} \le \int_\varepsilon^\delta \frac{ds}{H^{-1}(F(\sigma s))} = \frac{1}{\sigma} \int_{\varepsilon\sigma}^{\delta\sigma} \frac{dt}{H^{-1}(F(t))}$$

and the first part now follows by letting $\varepsilon \to 0$ and applying (1.1.7).

On the other hand, for the second part of the lemma it is enough to consider only values $\sigma > 1$. Then, for small $\varepsilon > 0$, we have by Lemma 4.1.1 (i),

$$\int_{\varepsilon/\sigma}^{\delta/\sigma} \frac{ds}{H^{-1}(\sigma F(s))} \ge \int_{\varepsilon/\sigma}^{\delta/\sigma} \frac{ds}{H^{-1}(F(\sigma s))} = \frac{1}{\sigma} \int_\varepsilon^\delta \frac{dt}{H^{-1}(F(t))}.$$

Letting $\varepsilon \to 0$ and applying (1.1.5) gives the second result. □

Lemma 4.1.3. *Let $T > 0$ and assume*

$$q \in C(0,T), \qquad q > 0 \quad in \ (0,T). \tag{4.1.4}$$

Then every classical distribution solution $w = w(t)$ of the problem ($' = d/dt$)

$$\begin{cases} [\text{sign } w(t)] \cdot [q(t)\Phi(w'(t))]' \ge 0 & in \ (0,T), \\ w(0) = 0, \quad w(T) = m > 0 \end{cases} \tag{4.1.5}$$

is such that

$$w \ge 0, \quad w' \ge 0 \quad in \ (0,T). \tag{4.1.6}$$

Further, there exists $t_0 \in [0,T)$ with the property that

$$w \equiv 0 \quad in \ [0,t_0]; \qquad w > 0, \quad w' > 0 \quad in \ (t_0,T). \tag{4.1.7}$$

Proof. We first claim that $w \ge 0$ in $[0,T]$. If the conclusion fails, there would be t_0 and t_1, with $0 \le t_0 < t_1 < T$ such that $w(t_0) = w(t_1) = 0$ and $w < 0$ in (t_0,t_1). Then, multiplying (4.1.5) by w and integrating on $[t_0,t_1]$

yields by integration by parts (or simply by the distribution meaning of solutions with the test function $w(t)$ on $[t_0, t_1]$)

$$\int_{t_0}^{t_1} q(t)\Phi(w'(t))w'(t)dt \leq 0,$$

where the integrand is non-negative by (4.1.4) and the fact that $t\Phi(t) > 0$ for $t \neq 0$. That is, necessarily $w' \equiv 0$ on $[t_0, t_1]$. Hence $w \equiv 0$ on $[t_0, t_1]$, since $w(t_0) = w(t_1) = 0$. This contradiction proves the claim.

Define the set $J = \{t \in (0, T) : w'(t) > 0\}$. Then, obviously, $J \neq \emptyset$, since $w(0) = 0$ and $w(T) > 0$, while also J is open in $(0, T)$ since $w \in C^1(0, T)$. Let $t_0 = \inf J$, so $t_0 \in [0, T)$ and $w \equiv 0$ in $[0, t_0]$, since we already know that $w \geq 0$ in $[0, T]$. Now, for any fixed $t \in (t_0, T)$ there obviously exists $t_1 \in (t_0, t)$ such that $w'(t_1) > 0$. By integration of (4.1.5) on $[t_1, t]$, recalling that $w \geq 0$ on $(0, T)$, we get

$$q(t)\Phi(w'(t)) \geq q(t_1)\Phi(w'(t_1)) > 0$$

by (4.1.4) and $(A2)$, so that $w' > 0$ on $(t_0, T]$. In turn, by integration, $w > 0$ in (t_0, T), proving (4.1.7). $\qquad\square$

Lemma 4.1.4. *If in Lemma 4.1.3 the hypothesis (4.1.4) is strengthened to*

$$q \in C(0, T), \qquad q > 0 \quad in \ (0, T), \qquad q \ non\text{-}increasing,$$

then w is convex on $[0, T]$ and

$$0 \leq w'(0) \leq m/T. \tag{4.1.8}$$

Proof. Indeed from (4.1.5) and (4.1.6) it follows that $q(t)\Phi(w'(t))$ is non-decreasing, and then since $q(t)$ is positive and non-increasing also $\Phi(w'(t))$ is non-decreasing. But Φ is increasing, so w' is non-decreasing. In turn, w is convex on $[0, T]$ and then (4.1.8) follows at once since $w(T) = m$. $\qquad\square$

Lemma 4.1.5. *Assume*

$$q \in C[0, T), \qquad q > 0 \quad in \ (0, T).$$

Then along every classical distribution solution w of the problem

$$\begin{cases} [q(t)\Phi(w'(t))]' - q(t)f(w(t)) \leq 0 & in \ (0, T), \\ w(0) = 0; \quad 0 \leq w \leq \delta, \quad w' \geq 0 & in \ (0, T), \end{cases} \tag{4.1.9}$$

there holds

$$\Phi(w'(t)) \le \frac{f(w(t))}{q(t)} \int_0^t q(s)\,ds + \frac{q(0)}{q(t)}\Phi(w'(0^+)), \quad t \in (0,T), \quad (4.1.10)$$

where $w'(0+)$ is defined as $\limsup_{t \to 0^+} w'(t)$.
    In particular, if $w'(0) = 0$, then (4.1.10) reduces to

$$\Phi(w'(t)) \le \frac{f(w(t))}{q(t)} \int_0^t q(s)\,ds. \qquad (4.1.11)$$

Proof. Integrating (4.1.9) on $[\tau, t]$, with $0 < \tau < t < T$, yields

$$q(t)\Phi(w'(t)) - q(\tau)\Phi(w'(\tau)) \le \int_0^t q(s)f(w(s))ds, \qquad (4.1.12)$$

and (4.1.10) follows at once by (F2), i.e., $f(w(s)) \le f(w(t))$ since $0 \le w(s) \le w(t) < \delta$, together with the $\limsup$ as $\tau \to 0^+$. $\square$

Lemma 4.1.6. *Assume*

$$q \in C^1[0,T] \quad and \quad q > 0 \quad in \ [0,T). \qquad (4.1.13)$$

Then along every classical distribution solution $w \in C^1(0,T)$ of the problem (4.1.9) for which $w'(0) = 0$ and the condition

$$\Phi \circ w' \in C^1(0,T) \qquad (4.1.14)$$

is satisfied,[1] we have

$$H(w'(t)) \le B(t)F(w(t)), \qquad t \in (0,T), \qquad (4.1.15)$$

where

$$B(t) = 1 + \sup_{s \in (0,t)} \left(-\frac{q'(s)}{q(s)^2} \int_0^s q(\tau)d\tau \right)^+. \qquad (4.1.16)$$

Furthermore, if $q' \ge 0$, then (4.1.15) becomes $H(w'(t)) \le F(w(t))$.

---

[1] For the main application of this lemma in Section 4.2 this condition holds without any difficulty; see (4.2.3) in Proposition 4.2.1.

Proof. Since $\Phi(w') \in C^1(0, T)$ by assumption, so also $H(w') \in C^1(0, T)$ by Lemma 4.1.1 (ii). Then by (4.1.2) and (4.1.9) one finds for $t \in (0, T)$ that

$$
\begin{aligned}
\{H(w'(t))\}' &= [\Phi(w'(t))]'w'(t) \\
&\le -\frac{q'(t)}{q(t)}\Phi(w'(t))w'(t) + f(w(t))w'(t),
\end{aligned}
\tag{4.1.17}
$$

since by assumption $w' \ge 0$, $q > 0$ in $(0, T)$. Integrating (4.1.17) on $(0, t)$, with $0 < t < T$, yields

$$
\begin{aligned}
H(w'(t)) &\le F(w(t)) - \int_0^t \frac{q'(s)}{q(s)}\Phi(w'(s))w'(s)\,ds \qquad \text{(since } w'(0) = 0\text{)}, \\
&\le F(w(t)) + \int_0^t \left(-\frac{q'(s)}{q(s)^2}\int_0^s q(\tau)\,d\tau\right)^+ f(w(s))w'(s)\,ds \\
&\le B(t)F(w(t))
\end{aligned}
$$

by (4.1.11) and (4.1.16). $\square$

Proposition 4.1.7. *Assume* (4.1.13). *Let w be a classical distribution solution of the problem*

$$
\begin{cases}
[q(t)\Phi(w'(t))]' - q(t)f(w(t)) \le 0 & \text{in } (0, T), \\
w(0) = 0, \quad w(T) = m > 0, & w' \ge 0,
\end{cases}
\tag{4.1.18}
$$

for which (4.1.14) *is satisfied. Suppose that $f(u) > 0$ for $u > 0$. If $w'(0) = 0$, then*

$$
\int_{0+} \frac{ds}{H^{-1}(F(s))} < \infty.
\tag{4.1.19}
$$

Proof. From the second line of (4.1.18) it is evident that there exists $t_0 \in [0, T)$ such that $w(t) = 0$ for $0 \le t \le t_0$ while $w > 0$ in (t_0, T). If $t_0 = 0$, then $w'(0) = 0$ by hypothesis, while if $t_0 > 0$, then in turn $w(t_0) = w'(t_0) = 0$ since $w \in C^1(0, T)$.

Let $t_2 \in (t_0, T)$. Clearly there exists $t_1 \in (t_0, t_2)$ such that $m_1 = w(t_1) > 0$ satisfies

$$
m_1 < \delta/B, \qquad F(Bm_1) < H(\infty),
$$

where $B = B(t_2)$ is given in Lemma 4.1.6. From this lemma applied to the interval (t_0, t_1), we thus obtain (see (4.1.15))

$$
H(w'(t)) \le B(t)F(w(t)) \le BF(w(t)) \quad \text{in } (t_0, t_1).
$$

since $B(t)$ is obviously non-decreasing. In turn by Lemma 4.1.1 (i), with $\sigma = 1/B$,

$$H(w'(t)) \leq F(Bw(t)) \quad \text{in } (t_0, t_1),$$

that is $w > 0$, $w'(t) \leq H^{-1}(F(Bw(t)))$ on (t_0, t_1). Using the fact that $f(u) > 0$ for $u > 0$ (and so also $F(u) > 0$ for $u > 0$), integration now yields, by the change of variables $s = Bw(t)$,

$$\int_0^{Bm_1} \frac{ds}{H^{-1}(F(s))} = B \int_{t_0}^{t_1} \frac{w'(t)dt}{H^{-1}(F(Bw(t)))} \leq B(t_1 - t_0) < \infty,$$

as required. $\qquad\qquad\qquad\qquad\qquad\qquad\qquad\qquad\qquad\qquad\qquad\qquad \square$

4.2 Existence theorems

In this section we shall obtain existence and uniqueness theorems for the boundary value problem

$$\begin{cases} [q(t)\Phi(w'(t))]' - q(t)f(w(t)) = 0 & \text{in } (0, T), \\ w(0) = 0, \quad w(T) = m > 0. \end{cases} \tag{4.2.1}$$

The main existence theorem, Proposition 4.2.1, will be used to obtain (radial) comparison functions for the proofs in later sections and in Chapter 5. Importantly here, we are able to use a weakened version of condition (F2), namely

(F3) $f(0) = 0$ and f is non-negative on some interval $[0, \delta)$, with δ possibly infinite.

Accordingly *it will be assumed in both Propositions 4.2.1 and 4.2.2 that* $m \in (0, \delta)$. Of course, in addition to (F3), conditions (A1), (A2), (F1) will be maintained throughout the section.

We suppose that the function q in (4.2.1) is continuous with $q > 0$ in $[0, T]$. Put

$$q_0 = \min_{[0,T]} q(t) > 0, \qquad q_1 = \max_{[0,T]} q(t) > 0. \tag{4.2.2}$$

Proposition 4.2.1.

(i) *Let* $\Phi(\infty) = \infty$. *Then problem* (4.2.1) *admits a classical distribution solution with the properties*

$$w \in C^1[0, T], \quad \Phi(w') \in C^1[0, T]; \quad w' \geq 0. \tag{4.2.3}$$

Moreover, for any such solution of (4.2.1) we have $w'(T) > 0$ and

$$\|w'\|_\infty \le \Phi^{-1}\left(\frac{q_1}{q_0}\left[T\bar{f}(m) + \Phi(m/T)\right]\right), \qquad (4.2.4)$$

where $\bar{f}(m) = \max_{u \in [0,m]} f(u)$. In particular, $w' \le 1$ if m is suffi-ciently small.

(ii) *Suppose $\Phi(\infty) = \omega < \infty$. Let $m \in (0, \delta)$ be such that*

$$\frac{q_1}{q_0}\left[T\bar{f}(m) + \Phi(m/T)\right] < \omega. \qquad (4.2.5)$$

Then the conclusion of part (i) continues to hold.

The proof relies on an application of the Leray–Schauder theorem to a carefully chosen homotopy $\mathcal{H} : X \times [0,1] \to X$, $X = (C[0,T], \|\cdot\|_\infty)$, defined by

$$\mathcal{H}[w,\sigma](t) = \sigma m - \int_t^T \Phi^{-1}\left(\frac{1}{q(s)}\left[\mu_\sigma - \sigma\int_s^T q(\tau)f(w(\tau))d\tau\right]\right) ds,$$

where $\mu_\sigma = \mu(w,\sigma)$ is the unique number such that

$$\mathcal{H}[w,\sigma](0) = 0,$$

so the mapping $\mathcal{H}[w,\sigma]$ is well defined. For the proof of Proposition 4.2.1 we refer the reader to the Appendix, Section 4.5; see also [81, Proposition 4.1].

The condition $f(0) = 0$ in (F3) is crucial for Proposition 4.2.1. In fact the proposition fails otherwise, as shown by the example $f(u) \equiv 1$, $q \equiv 1$, and $A(s) \equiv 1$. In this case every non-negative solution of (4.2.1) must have the form $w(t) = at + \frac{1}{2}t^2$, $a \ge 0$, which gives the extraneous condition for solvability $m = w(T) = aT + \frac{1}{2}T^2 \ge \frac{1}{2}T^2$.

In view of (4.2.3) we note that, for the given solution w, all derivatives with respect to t in (4.2.1) can equally well be understood as ordinary derivatives, no recourse to distribution solutions in fact being needed.

Proposition 4.2.2. *Let $q \in C[0,T] \cap C^1[0,T)$ with $q > 0$ in $[0,T]$. Suppose that (F2) is satisfied and that either $f = 0$ in $u \in [0,d]$, $d > 0$, or that (1.1.5) holds, that is*

$$\int_{0+} \frac{ds}{H^{-1}(F(s))} = \infty. \qquad (4.2.6)$$

Then the solution of (4.2.1) *given in Proposition 4.2.1 has the properties*

$$w > 0 \quad in \ (0, T], \qquad w' > 0 \quad in \ [0, T]. \tag{4.2.7}$$

Proof. Case 1. Let $f = 0$ in $[0, \mu]$. Then from (4.2.1) we have $[q(t)\Phi(w'(t))]'$ $= 0$ at least for t near 0. Hence in turn $q\Phi \circ w' = $ Constant > 0 for small t (if the constant is zero, then $w' = 0$ for small $t > 0$, and then by continuation for all $t > 0$, which contradicts the boundary condition $w = m$ at $t = T$). Consequently $w'(0) = \Phi^{-1}(\text{Constant}/q(0)) > 0$, so from Lemma 4.1.3 and the fact that $t_0 = 0$ in the present case, we get $w'(t) > 0$ in $[0, T]$ and $w > 0$ in $(0, T]$ as required.

Case 2. Let (4.2.6) hold. Note that (4.1.14) is satisfied in view of (4.2.3). Also we already know that $w'(0) \geq 0$ and $0 \leq w \leq m$. In fact, the case $w'(0) = 0$ cannot occur by Proposition 4.1.7 and assumption (4.2.6). Consequently $w'(0) > 0$ and the required conclusion then follows as before. □

Propositions 4.2.1 and 4.2.2 have the following useful corollary, which later will take the role of Lemma 2.8.2 for divergence structure equations.

Lemma 4.2.3. *Let B_R be an arbitrary open ball of radius R in $\mathbb{R}^n$ and let $E_R = B_R \setminus \overline{B_{R/2}}$. If $\Phi(\infty) = \infty$, then for every $m \in (0, \delta)$ there exists a non-negative function $v \in C^1(\overline{E_R})$ which is a solution of*

$$\text{div}\{A(|Dv|)Dv\} - f(v) = 0 \tag{4.2.8}$$

in the annulus E_R, with boundary values

$$v = 0 \quad on \ \partial B_R, \qquad v = m \quad on \ \partial B_{R/2}. \tag{4.2.9}$$

If $\Phi(\infty) = \omega < \infty$, the conclusion remains valid provided $m \in (0, \delta)$ is so small that

$$2^{n-1}[R\bar{f}(m)/2 + \Phi(2m/R)] < \omega, \tag{4.2.10}$$

where $\bar{f}(m) = \max_{u \in [0,m]} f(u)$.

Furthermore, if (F2) *holds and either $f \equiv 0$ in $[0, d]$, $d > 0$, or (4.2.6) is satisfied, then $|Dv| > 0$ in $\overline{E_R}$ and in particular $\partial_\nu v < 0$ on ∂B_R, where ν is the exterior unit normal to B_R.*

Proof. In Proposition 4.2.1 choose $q(t) = (R - t)^{n-1} = r^{n-1}$, $t \in [0, R/2]$, and $v(x) = w(t)$, $t = R - r$, $r = |x - x_0|$, where x_0 denotes the center of B_R. Then v is a radial solution of (4.2.8)–(4.2.9) in the annulus E_R. The final part of the lemma follows from Proposition 4.2.2. □

As a particular example, an existence theorem for the problem

$$\begin{cases} [q(t)\Phi(w'(t))]' - a(t)q(t)f(w(t)) = h(t) & \text{in } (0,T), \\ w(0) = 0, \quad w(T) = m > 0 \end{cases} \tag{4.2.11}$$

can be given, exactly following the ideas of Proposition 4.2.1. The result is stated in

Proposition 4.2.4. *Assume a, h, $q \in C[0,T]$ and $h \geq 0$, $a \geq 0$, $q > 0$ in $[0,T]$. Then problem (4.2.11) with $m \in (0,\delta)$, and with m and $\int_0^T h(t)\,dt$ suitably small in case $\Phi(\infty) < \infty$, admits a classical distribution solution with the properties $w \in C^1[0,T]$, $w' \geq 0$.*

The question of uniqueness of solutions of (4.2.1) and (4.2.11) is also of interest. For this result, we assume the main conditions (A1), (A2), (F1), (F2).

Theorem 4.2.5. *Assume $a, h, q \in C(0,T)$ and $a \geq 0$, $q > 0$ in $(0,T)$. Then problems (4.2.1) and (4.2.11) admit at most one classical distribution solution with range in $[0,\delta)$.*

Proof. Let w and $\tilde{w}$ be two solutions of (4.2.11) with ranges in $[0,\delta)$. Then by (4.2.11) together with (A2) and (F2), we obtain

$$0 \leq \int_0^T q(t)[\Phi(w'(t)) - \Phi(\tilde{w}'(t))] \cdot [w'(t) - \tilde{w}'(t)]dt$$

$$= -\int_0^T a(t)q(t)[f(w(t)) - f(\tilde{w}(t))] \cdot [w(t) - \tilde{w}(t)]dt \leq 0.$$

It now follows at once that $w \equiv \tilde{w}$ in $[0,T]$ since Φ is strictly increasing. $\square$

4.3 Existence theorems on a half-line

In the next section we shall prove the necessity part of Theorem 1.1.2 through the existence of classical solutions of the exterior Dirichlet problem for (1.1.2), with equality sign. Because of the separate and independent interest of this question, we devote the present section to its consideration.

For the following main theorem we maintain conditions (A1), (A2), (F1), and in place of (F2) the slightly stronger condition

(F2)′ $f(0) = 0$ and f is positive and non-decreasing in $(0,\delta)$, $\delta > 0$ finite.

Clearly (F2)′ implies (F2) which implies (F3).

Theorem 4.3.1 (Exterior Dirichlet Problem). *For all $R > 0$ and $m \in (0, \delta)$, with m sufficiently small if $\Phi(\infty) = w < \infty$, there is a classical C^1 radial solution $u(x) = u(r)$ of the problem*

$$\operatorname{div}\{A(|Du|)Du\} - f(u) = 0, \qquad u \geq 0 \qquad (4.3.1)$$

in $\Omega_R = \{x \in \mathbb{R}^n \; : \; |x| > R\}$, such that

$$u(R) = m, \qquad u(x) \to 0 \quad as \; |x| \to \infty. \qquad (4.3.2)$$

Moreover $u' < 0$ whenever $u > 0$.

The required smallness condition on m when $w < \infty$ is given below by (4.3.6).

Proof. First consider the case when $w = \infty$. Let $j = 1, 2, \ldots$, define $q(t) = (R + j - t)^{n-1}$ and denote by w_j the solution of

$$\begin{cases} [q(t)\Phi(w_t(t))]_t - q(t)f(w(t)) = 0, \\ w(0) = 0, \qquad w(j) = m \in (0, \delta), \\ w_t \geq 0 \quad \text{in } [0, j], \end{cases}$$

which exists by Proposition 4.2.1. Moreover because q is positive and decreasing, then w is convex by Lemma 4.1.4.

It follows now that $u_j(r) = w_j(t)$, $t = R + j - r$, is a solution of

$$\begin{cases} [r^{n-1}\Phi(u'(r))]' - r^{n-1}f(u(r)) = 0 \quad (' = d/dr), \\ u(R) = m, \qquad u(R + j) = 0, \\ u' \leq 0 \quad \text{in } [R, R + j] \end{cases} \qquad (4.3.3)$$

(here recall that Φ is defined for negative s, according to the agreement at the beginning of Section 4.1, namely $\Phi(s) = -\Phi(-s)$). It is obvious that the equation in (4.3.3) is exactly the radial version of (4.3.1).

We claim that the sequence $j \mapsto u_j$ is non-decreasing. Indeed, u_j and u_{j+1} are C^1 radial solutions of $\operatorname{div}\{A(|Du|)Du\} - f(u) = 0$ in the annulus $\mathscr{A}_j = \{x \in \Omega_R \; : \; R \leq |x| \leq R + j\}$. Obviously $u_j \leq u_{j+1}$ on $\partial\mathscr{A}_j$ so that $u_j \leq u_{j+1}$ in $\mathscr{A}_j$ by (F2)', Theorem 2.4.1 and Proposition 2.4.2, as claimed.[2]

---

[2]It is of interest that the monotonicity of the sequence $(u_j)_j$ can be obtained under the weaker condition (F3) instead of (F2)'. More specifically, under (F3) monotonicity follows as in the proof of Theorem 3.6.4 of [39]. Since the main application of Theorem 4.3.1 in Chapter 5 deals with nonlinearities f satisfying (F2), we shall not pursue this further.

Each u_j is continuous, non-increasing and non-negative in $[R, R+j]$. Hence by the Dini theorem the sequence $(u_j)_j$ converges uniformly on every compact subset of $[R, \infty)$ to the non-negative, non-increasing, continuous limit u. We shall show that u is the required radial solution of (4.3.1), (4.3.2). Of course $u : [R, \infty) \to [0, m]$, with $u(R) = m$.

In fact, corresponding to (4.3.3), the function u_j satisfies the integral equation on $[R, R+j]$,

$$u_j(r) = m - \int_R^r \Phi^{-1}\left(s^{1-n}\left[\mu_j - \int_R^s \tau^{n-1}f(u_j(\tau))d\tau\right]\right) ds,$$

where μ_j is determined by the condition $u_j'(R) = -\Phi^{-1}(R^{1-n}\mu_j)$. In other words

$$\mu_j = R^{n-1}\Phi(|u_j'(R)|) > 0$$

since $|u_j'(R)| \le |u_1'(R)|$ by monotonicity and the fact that $u_j(R) = m$ for each j. The positive non-increasing sequence $(\mu_j)_j$ converges to some number $\mu \ge 0$. Letting $j \to \infty$ the limit function u satisfies the integral equation

$$u(r) = m - \int_R^r \Phi^{-1}\left(s^{1-n}\left[\mu - \int_R^s \tau^{n-1}f(u(\tau))d\tau\right]\right) ds. \qquad (4.3.4)$$

By the continuity of u in $[R, \infty)$ it follows from (4.3.4) that u is of class $C^1[R, \infty)$. Thus u is also a classical distribution solution of

$$\begin{cases} [r^{n-1}\Phi(u'(r))]' - r^{n-1}f(u(r)) = 0 & \text{in } [R, \infty), \\ u(R) = m; \quad u \ge 0, \quad u' \le 0 & \text{in } [R, \infty). \end{cases} \qquad (4.3.5)$$

Of course, the equation on the first line of (4.3.5) is equivalent to (4.3.1) for radial functions $u = u(r)$.

To complete the proof of the theorem in the case $\omega = \infty$, it remains to show that $u' < 0$ when $u > 0$ and that $u(r) \to 0$ as $r \to \infty$. To obtain the first, note by virtue of (4.3.5) that should $u' = 0$ at some point r_0 where $u > 0$, then by (F2)' we would have $r^{n-1}\Phi(u'(r)) > 0$ for all $r \in (r_0, r_0+\varepsilon)$, $\varepsilon > 0$ sufficiently small; that is $u'(r) > 0$ for $r \in (r_0, r_0+\varepsilon)$, which is absurd. Hence $u' < 0$ at any point where $u > 0$.

Denote by ℓ the non-negative finite limit of u as $r \to \infty$. Since u' is non-decreasing by convexity, then $u'(r) \to 0$ as $r \to \infty$. Integrating (4.3.5) on $[r, r+1]$, with $R \leq r < \infty$, we get

$$\Phi(u'(r+1)) - \left(\frac{r}{r+1}\right)^{n-1} \Phi(u'(r)) = \frac{1}{(r+1)^{n-1}} \int_r^{r+1} \tau^{n-1} f(u(\tau)) \, d\tau$$

$$\geq \left(\frac{r}{r+1}\right)^{n-1} f(\ell)$$

by (F2) and the fact that $\ell \leq u(r) \leq \delta$ for $r \in [R, \infty)$. Letting $r \to \infty$ then yields $0 \geq f(\ell) \geq 0$, that is $\ell = 0$ by (F2)'. This completes the proof when $\omega = \infty$.

We now treat the case when $\omega < \infty$. Suppose $m < \delta$ so small that

$$\left(\frac{R+1}{R}\right)^{n-1} [f(m) + \Phi(m)] = \hat{\omega} < \omega. \tag{4.3.6}$$

We introduce a new operator $\hat{\Phi}$, defined by

$$\hat{\Phi}(s) = \begin{cases} \Phi(s) & \text{for } 0 \leq s \leq \Phi^{-1}(\hat{\omega}), \\ \dfrac{\hat{\omega}}{\Phi^{-1}(\hat{\omega})} s & \text{for} \quad s \geq \Phi^{-1}(\hat{\omega}). \end{cases} \tag{4.3.7}$$

Clearly $\hat{\Phi}$ is continuous and increasing on $[0, \infty)$, thus satisfying (A1) and (A2), and moreover $\hat{\Phi}(\infty) = \infty$.

We apply the first part of the proof with Φ replaced by $\hat{\Phi}$ and u replaced by $\hat{u}$. Clearly the solution $\hat{u}$ exists. It will be a solution of the original problem (4.3.1), (4.3.2), provided $\|\hat{u}'\|_\infty \leq \Phi^{-1}(\hat{\omega})$. But by convexity, (4.2.4), and (4.3.6),

$$\|\hat{u}'\|_\infty \leq |\hat{u}'(R)| \leq |\hat{u}_1'(R)| \leq \hat{\Phi}^{-1}\left(\frac{q_1}{q_0}[f(m) + \Phi(m)]\right)$$

$$= \hat{\Phi}^{-1}(\hat{\omega}) = \Phi^{-1}(\hat{\omega}),$$

since $q_1/q_0 = [(R+1)/R]^{n-1}$. This completes the proof. $\square$

The solution $u = u(r)$ given by Theorem 4.3.1 is unique, the precise result being

Theorem 4.3.2. *Let the hypotheses of Theorem 4.3.1 be satisfied. There cannot be more than one solution of (4.3.1), (4.3.2) in Ω_R, whether radial or not, which has range in $[0, \delta)$.*

This is an immediate consequence of the comparison Theorem 2.4.1 together with Proposition 2.4.2.

Theorem 4.3.3. *Let the hypotheses of Theorem 4.3.1 be satisfied. Then the solution u given by Theorem 4.3.1 is everywhere positive provided that (1.1.5) holds. Conversely if (1.1.7) is satisfied, then u has compact support.*

Theorem 4.3.3 will be proved in Chapter 5.

Remark. When $\omega < \infty$ the condition (4.3.6) is not best possible, and can be improved to the form

$$T_0 f(m) + \Phi\left(\frac{m}{T_0}\right) < \left(\frac{R}{R+T_0}\right)^{n-1} \omega,$$

where $T_0 > 0$ is a positive parameter which can be assigned arbitrarily; this follows easily by replacing the interval $[R, R+1]$ in the definition of u_1 by $[R, R+T_0]$ for any $T_0 > 0$.

As an example, when $R \ll 1$ and $A(s) = 1/\sqrt{1+s^2}$ is the mean curvature operator, with $f(z) = \kappa z$, $\kappa > 0$, and $n \geq 2$ (equation of a capillary surface under gravity), by taking $T_0 = aR$ with $a > 0$ suitably small, we get the solvability condition $m < R/(n-1)$; whereas from (4.3.6) one gets the alternative condition $m < R/\kappa$.

A different approach to the radial exterior problem, containing a number of precise estimates in the case when $\omega < \infty$ and $\Phi'(0) > 0$, has been given by Turkington [112].

4.4   The end point lemma

In this section we prove a remarkable result having important consequences for the Strong Maximum Principle in Chapter 5 and the phenomenon of dead cores in Section 8.4. In what follows we maintain the conditions (A1), (A2), (F1) and (F2).

Lemma 4.4.1 (End Point Lemma). *Suppose $f(u) > 0$ for $u > 0$ and that (1.1.7) is satisfied. For fixed $\sigma > 0$, define*

$$C_\sigma = \int_0^\delta \frac{ds}{H^{-1}(\sigma F(s))}. \tag{4.4.1}$$

Then for every $C \in (0, C_\sigma)$ there exists a number $\gamma = \gamma(C) \in (0, \delta)$ and a function $w \in C^1[0, C]$ such that

(i) $\gamma \to 0 \quad$ *as* $C \to 0$,

(ii) $w(0) = w'(0) = 0$, $w(C) = \gamma$; $0 \leq w' \leq H^{-1}(F(\gamma))$,

(iii) $\qquad\qquad\qquad [\Phi(w'(t))]' = \sigma f(w(t)) \quad for\ t \in (0, C),$

(iv) $\qquad\qquad\qquad \Phi(w'(t)) \leq \sigma t f(w(t)) \quad for\ t \in (0, C).$

[*If $H(\infty)$ is finite we take $\delta > 0$ so small that $\sigma F(\delta) < H(\infty)$.*]

Proof. First note that the integral in (4.4.1) is convergent, in view of Lemma 4.1.2 and (1.1.7). For given $C \in (0, C_\sigma)$, we take $\gamma \in (0, \delta)$ so that

$$0 < C = \int_0^\gamma \frac{ds}{H^{-1}(\sigma F(s))};$$

clearly $\gamma = \gamma(C)$ is uniquely determined by C, and of course $\gamma \to 0$ as $C \to 0$.

Now define $w : [0, C] \to \mathbb{R}$ by

$$t = \int_0^{w(t)} \frac{ds}{H^{-1}(\sigma F(s))}. \tag{4.4.2}$$

Hence

$$\frac{w'(t)}{H^{-1}(\sigma F(w(t)))} = 1, \qquad 0 < t < C,$$

that is $H(w') = \sigma F(w)$. Thus in turn $[H(w')]' = \sigma f(w)w'$. Obviously part (ii) of the lemma is satisfied; moreover, since $w' > 0$ on $(0, C]$, from Lemma 4.1.1 (ii) we obtain part (iii).

An integration using parts (ii), (iii) and the monotonicity of f in (F2)$'$ shows that also $\Phi(w'(t)) \leq \sigma t f(w(t))$. This completes the proof. $\qquad\square$

There is a slightly stronger result, proved in the same way.

Lemma 4.4.2. *The conclusions* (i)–(iii) *of Lemma 4.4.1 are valid if condition* (F2), *together with the positivity of f, is replaced by the weaker condition that $f(0) = 0$ and F is positive in some interval $(0, \delta)$.*

4.5 Appendix: Proof of Proposition 4.2.1

For the proof of this proposition only, we shall redefine the operator Φ in $\mathbb{R}^-$ by setting $\Phi(s) = s$ when $s < 0$; this can be done without loss of generality since the ultimate solution w satisfies $w' \geq 0$.

Case (i). Let

$$\mu_1 = q_1[T\bar{f}(m) + \Phi(m/T)], \quad I = [0, \mu_1], \quad \bar{f}(m) = \max_{u \in [0,m]} f(u). \tag{4.5.1}$$

It is convenient also to redefine f so that $f(z) = f(m)$ for all $z \geq m$. This will not affect the conclusion of the proposition, since clearly any ultimate solution with $w' \geq 0$ satisfies $0 \leq w \leq m$. We recall the earlier agreement that $f(z) = 0$ for $z \leq 0$.

With these preliminaries settled, we can proceed to the main proof. We shall make use of the Leray–Schauder fixed point theorem, an idea suggested in this context by Montenegro.

Denote by X the Banach space $C[0, T]$, endowed with the usual norm $\| \cdot \|_\infty$, and let $\mathcal{T}$ be the mapping from X to X defined for $t \in [0, T]$ by

$$\mathcal{T}[w](t) = m - \int_t^T \Phi^{-1}\left(\frac{1}{q(s)}\left[\mu - \int_s^T q(\tau)f(w(\tau))d\tau \right] \right) ds, \quad (4.5.2)$$

where $\mu = \mu(w) \in I$ is chosen so that

$$\mathcal{T}[w](0) = 0. \qquad (4.5.3)$$

We shall show that such a choice of μ is uniquely possible.

Indeed for any fixed $w \in X$ and for any $\mu \in I$ we have

$$-\frac{\bar{f}(m)}{q_0} \int_0^T q(t)\, dt \leq \frac{1}{q(s)}\left[\mu - \int_s^T q(\tau)f(w(\tau))d\tau \right] \leq \frac{\mu_1}{q_0}. \qquad (4.5.4)$$

It follows now that $\mathcal{T}[w]$ is well defined for each fixed μ in I.

Moreover for $\mu = 0$ we see that, for all $w \in X$,

$$\mathcal{T}[w](0) \geq m.$$

On the other hand, for $\mu = \mu_1$ we find, for all w in X,

$$\mathcal{T}[w](0)$$
$$= m - \int_0^T \Phi^{-1}\left(\frac{q_1}{q(s)}\Phi(m/T) + \frac{1}{q(s)}\left[q_1 T\bar{f}(m) - \int_s^T q(\tau)f(w(\tau))d\tau \right] \right) ds$$
$$\leq m - \int_0^T \Phi^{-1}(\Phi(m/T))ds = 0,$$

where we have used the condition (4.5.1), the definition of q_1 in (4.2.2), and the fact that $0 \leq f(z) \leq \bar{f}(m)$. Since the integral on the right side of (4.5.2) is a strictly increasing function of μ for fixed w, it is now obvious that there exists a unique $\mu \in I$ such that (4.5.3) holds.

Define the homotopy $\mathcal{H} : X \times [0, 1] \to X$ by

$$\mathcal{H}[w, \sigma](t)$$
$$= \sigma m - \int_t^T \Phi^{-1} \left(\frac{1}{q(s)} \left[\mu_\sigma - \sigma \int_s^T q(\tau) f(w(\tau)) d\tau \right] \right) ds, \qquad (4.5.5)$$

where $\mu_\sigma = \mu(w, \sigma) \in I$ is a number chosen such that

$$\mathcal{H}[w, \sigma](0) = 0.$$

Clearly, as in the case of the mapping $\mathcal{T}$ in (4.5.2), such a value μ_σ exists and is unique, and the mapping $\mathcal{H}[w, \sigma]$ is accordingly well defined.[3]

By construction, any fixed point $w_\sigma = \mathcal{H}[w_\sigma, \sigma]$ is of class $C^1[0, T]$, has the property that $\Phi(w') \in C^1[0, T]$, and is a classical distribution solution of the problem

$$\begin{cases} [q(t)\Phi(w'_\sigma(t))]' - \sigma q(t) f(w_\sigma(t)) = 0 & \text{in } [0, T], \\ w_\sigma(0) = 0, \quad w_\sigma(T) = \sigma m. \end{cases} \qquad (4.5.6)$$

Moreover, by Lemma 4.1.3, a fixed point $w = \mathcal{H}[w, 1]$ satisfies w, $w' \geq 0$, and so is a solution of problem (4.2.1) satisfying the conditions (4.2.3), with $w' \geq 0$.

It remains to show that such a fixed point $w = w_1$ exists. We shall use Browder's version of the Leray–Schauder theorem for this purpose (see Theorem 11.6 of [43]).

To begin with, obviously $\mu_\sigma = 0$ when $\sigma = 0$, and so $\mathcal{H}[w, 0](t) \equiv 0$ for all w in X, that is $\mathcal{H}[w, 0]$ maps X into the single point $w_0 = 0$ in X. (This is the first hypothesis required in the application of the Leray–Schauder theorem at the end of the proof.)

We show next that $\mathcal{H}$ is compact and continuous from $X \times [0, 1]$ into X. Let $(w_k, \sigma_k)_k$ be a bounded sequence in $X \times [0, 1]$. Clearly $\mu_{\sigma_k} \in I$; therefore again using the fact that $0 \leq f(z) \leq \bar{f}(m)$ for all $z \geq 0$, together with (4.5.4), it is clear that

$$\|\mathcal{H}'[w_k, \sigma_k]\|_\infty \leq C',$$

---

[3] The simpler homotopy

$$\tilde{\mathcal{H}}[w, \sigma](t) = \int_0^t \Phi^{-1} \left(\frac{1}{q(s)} \left[\kappa + \sigma \int_0^s q(\tau) f(w(\tau)) d\tau \right] \right) ds$$

with $\kappa = \kappa(w)$ chosen so that $\tilde{\mathcal{H}}[w, \sigma](T) = m$, is in fact less convenient in carrying out the proof.

where (recalling that $\Phi^{-1}(t) = t$ when $t < 0$)

$$C' = \max\left\{\frac{\bar{f}(m)}{q_0}\int_0^T q(t)dt, \ \Phi^{-1}(\mu_1/q_0)\right\}. \tag{4.5.7}$$

It is now an immediate consequence of the Ascoli–Arzelà theorem that $\mathcal{H}$ maps bounded sequences into relatively compact sequences in X.

We claim finally that $\mathcal{H}$ is continuous on $X \times [0, 1]$. Indeed, let $w_j \to w$, $\sigma_j \to \sigma$, $(w_j, \sigma_j) \in X \times [0, 1]$. Then in (4.5.5) clearly $\sigma_j f(w_j) \to \sigma f(w)$, since the modified function f is continuous on $\mathbb{R}$. It must then be shown that $\mu(w_j, \sigma_j) \to \mu(w, \sigma)$. To this end, suppose for contradiction that this fails. Then, for some subsequence, still called (w_j, σ_j), we should have

$$\mu(w_j, \sigma_j) \to \tilde{\mu} \neq \mu = \mu(w, \sigma).$$

In this case, from (4.5.3) one gets by subtraction

$$\int_0^T \left\{\Phi^{-1}\left(\frac{1}{q(s)}\left[\tilde{\mu} - \sigma\int_s^T q(\tau)f(w(\tau))d\tau\right]\right)\right.$$
$$\left. - \Phi^{-1}\left(\frac{1}{q(s)}\left[\mu - \sigma\int_s^T q(\tau)f(w(\tau))d\tau\right]\right)\right\} ds = 0. \tag{4.5.8}$$

But Φ^{-1} is a monotone increasing function of its argument, so clearly the integrand in (4.5.8) is either everywhere positive or everywhere negative, giving the required contradiction.

To apply the Leray–Schauder theorem it is now enough to show that there is a constant $M > 0$ such that

$$\|w\|_\infty \leq M \quad \text{for all } (w, \sigma) \in X \times [0, 1], \text{ with } \mathcal{H}[w, \sigma] = w. \tag{4.5.9}$$

Let (w, σ) be a pair of type (4.5.9). But, as observed above, since $w' \geq 0$, clearly $\|w\|_\infty = w(T) = \sigma m \leq m$. Thus we can take $M = m$ in (4.5.9).

The Leray–Schauder theorem therefore can be applied and the mapping $\mathcal{T}[w] = \mathcal{H}[w, 1]$ has a fixed point $w \in X$, which is the required solution of (4.2.1). That (4.2.3) holds for this solution was noted earlier in the proof.

The last part of the theorem is a direct consequence of (4.5.2) evaluated at a fixed point w, together with the right-hand inequality of (4.5.4) and the fact that $\mu \in I$.

Case (ii). The argument is exactly the same as before, with the single exception that in (4.5.4) the right-hand side μ_1/q_0 is now less than w by virtue of (4.2.10) and (4.5.1). Thus, T is well defined in X, and the rest of the proof is unchanged.   □

The proof of Proposition 4.2.4 goes in almost the same way, except one must take

$$\mu_1 = q_1[a_1 T \bar{f}(m) + \Phi(m/T)] + \int_0^T h(t)dt, \quad \text{where} \quad a_1 = \max_{t \in [0,T]} a(t),$$

rather than in (4.5.1).

Problems

4.1 Supply the details for the proof of Lemma 4.1.1.

4.2 Prove the existence Proposition 4.2.4 for problem (4.2.11), following the ideas of proof of Proposition 4.2.1.

4.3 Show that the monotonicity of the sequence $(u_j)_j$ in the proof of Theorem 4.3.1 can be obtained under the weaker condition (F3) instead of (F2)', following the proof of Theorem 3.6.4 of [39].

4.4 Supply the details for the proof of Theorem 4.3.2.

4.5 Ditto for Lemma 4.4.2.

Chapter 5

The Strong Maximum Principle and the Compact Support Principle

5.1 The strong maximum principle

With the work of the preceding Chapter 4 available, we can turn to the proofs of the Strong Maximum Principle, Theorem 1.1.1, and the Compact Support Principle, Theorem 1.1.2, stated in the Introduction.

Proof of Sufficiency in Theorem 1.1.1. We proceed exactly as in the proof of Hopf's maximum principle in Section 2.8, with the two exceptions that (a) the weak maximum principle, Theorem 2.8.1, is replaced by Theorem 2.4.1 and Proposition 2.4.2, and (b) Lemma 2.8.2 is replaced by Lemma 4.2.3. In particular the crucial condition (iii) of Lemma 2.8.2 is obtained from the last part of Lemma 4.2.3, in view of the key assumption (1.1.5). $\square$

As remarked in the introduction, the case of necessity in Theorem 1.1.1 is due to Diaz [28]. Theorem 1.1.1 is proved. $\square$

Another proof of the necessity part of (1.1.5). Suppose that $F > 0$ in some interval $(0, \delta)$, and that (1.1.5) fails, that is (1.1.7) holds. By the End Point Lemma 4.4.2 we can then introduce the function $w = w(t)$, of class $C^1[0, C]$, $C \in (0, C_1)$, $\sigma = 1$. Let $\Omega = \{x \in \mathbb{R}^n : x_n < C\}$ and define $u(x) = 0$ if $x_n \leq 0$, $u(x) = w(x_n)$ if $x_n \in [0, C)$. Hence $u \in C^1(\Omega)$ is non-negative by

Lemma 4.4.2 (ii), and is also a solution of (1.1.2), *with the equality sign,* by Lemma 4.4.2 (iii). Clearly $u(0) = w(0) = 0$ and at the same time $u \not\equiv 0$. Hence the strong maximum principle fails. □

Remarks. 1. The proof of sufficiency we have given is in fact not different in its underlying ideas from those in [10], [21], [30], [86], [113], the principal improvements here being the direct approach, the generality of the equation and the solution class, and the clarification of the method. The proof here uses only standard calculus, and the elementary Leray–Schauder theorem (see [43], Theorem 11.6), but requires neither monotone operator theory (as [113], [28]–[31]), nor Orlicz–Sobolev space theory, nor viscosity solution theory (as [49]), nor probabilistic methods.

We note also that Diaz, Saa and Thiel have stated a version of Theorem 1.1.1, see Theorem 6 of [31], but with partially insufficient proof.

2. The necessity of condition (1.1.5) for the Strong Maximum Principle can be obtained under a weaker hypothesis than (F2). In fact, it is enough to suppose only

$$f(0) = 0 \quad \text{and either} \quad f \equiv 0 \quad \text{or} \quad F(s) > 0 \quad \text{for } s \in (0, \delta).$$

This is because the principal construction required for Diaz' proof uses only this condition; see also the second proof of necessity given above.

3. The second proof of necessity for the Strong Maximum Principle also yields a direct and simple counterexample to the unique continuation question for the equation $\operatorname{div}\{A(|Du|)Du\} - f(u) = 0$, when (1.1.7) holds. That is, the function $u(x) = w(x_n)$ shows that a solution in a domain Ω may vanish in a subdomain without vanishing throughout Ω.

Proof of first part of Theorem 4.3.3. Because of (1.1.5) the strong maximum principle is valid for (1.1.2), hence also for (4.3.1). But since $u(R) = m > 0$ and because u is a non-negative (radial) solution of (4.3.1), it now follows that $u > 0$ on the entire domain of the solution. □

Example: the degenerate Laplacian. The strong maximum principle can be treated more simply in the case of the canonical p-Laplacian inequality, $p > 1$,

$$\Delta_p u - f(u) \leq 0, \qquad u \geq 0.$$

For our present purpose, we assume that

$$f(z) \leq cz^{p-1}, \tag{5.1.1}$$

the borderline case for (1.1.5).

An appropriate comparison function $v = v(r)$, $r = |x|$, can be taken in the form

$$v(r) = \alpha[(R/r)^\vartheta - 1], \qquad R/2 \le r \le R, \qquad (5.1.2)$$

where $\alpha = m/(2^\vartheta - 1)$ and ϑ, R are to be determined. Then

$$\Phi(|v'|) = |v'|^{p-1} = \left(\frac{\alpha\vartheta}{R}\right)^{p-1} \left(\frac{R}{r}\right)^{(p-1)(\vartheta+1)};$$

moreover, after a short calculation, there results

$$[|v'|^{p-1}]' + \frac{n-1}{r}|v'|^{p-1} + f(v)$$

$$\le (\alpha\vartheta)^{p-1} \left(\frac{R}{r}\right)^{(p-1)\vartheta} \left\{ \frac{n-1-(p-1)(\vartheta+1)}{r^p} + \frac{c}{\vartheta^{p-1}} \right\}.$$

This again will be ≤ 0 provided that

$$\vartheta = \frac{2(n-1)}{p-1} - 1, \qquad R \le \left[\frac{(n-1)}{c}\right]^{1/p} \vartheta^{1/p'}.$$

That is, $\Delta_p v - f(v) \ge 0$ for $R/2 \le |x| \le R$, and the proof of the strong maximum principle, Theorem 1.1.1, now applies unchanged, but avoiding the delicate arguments of Proposition 4.2.1, or of [113].

In summary, for the borderline case (5.1.1) of the p-Laplacian inequality, we get an elementary proof of Vázquez' strong maximum principle. At the same time, the simple comparison function (5.1.2) does not suffice for general operators or for more complicated nonlinearities. This observation indicates the need for the alternative deeper-lying construction of v in the proof of Theorem 1.1.1.

5.2 The compact support principle

Proof of sufficiency in Theorem 1.1.2. Let u be a (non-negative) solution of (1.1.6) in an exterior domain $\Omega \supset \Omega_R$ with $u(x) \to 0$ as $|x| \to \infty$. We must show that u has compact support in Ω. To begin with, clearly there exists $R_0 \ge R$ such that $u(x) \le \delta' < \delta$ if $|x| \ge R_0$. Let $w = w(t)$ be the function introduced in the alternative proof of the necessity part of Theorem 1.1.1, with $\sigma = 1$ and with C chosen so near C_1 that $\gamma(C) \ge \delta'$.

Define $\Omega_0 = \{x \in \mathbb{R}^n : |x| > R_0\}$ and $v(x) = w(C + R_0 - |x|)$ for $R_0 < |x| \leq C + R_0$. We extend the definition of v to all $x \in \Omega_0$ by setting $v(x) = 0$ when $|x| > C + R_0$. Clearly $v \in C^1(\Omega_0)$ by Lemma 4.4.2 (ii). Moreover, for $x \in \Omega_0$ and $r = |x|$, we have

$$\operatorname{div}\{A(|Dv|)Dv\} - f(v) = -[\Phi(|v'|)]' - \frac{(n-1)}{r}\Phi(|v'|) - f(v)$$

$$\leq [\Phi(w_t)]_t - f(w) \leq 0 \qquad (5.2.1)$$

in view of Lemma 4.4.2 (iii) and the fact that $\Phi(s) \geq 0$. Since $u(x) \leq \delta' \leq v(x)$ on $\partial\Omega_0$, and since $u(x), v(x) \to 0$ as $|x| \to \infty$, we can apply the comparison Theorem 2.4.1 and Proposition 2.4.2 to obtain $0 \leq u(x) \leq v(x)$ in Ω_0. In particular, $u(x) \equiv 0$ when $|x| \geq R_1 = R_0 + C$, as required. □

Proof of necessity in Theorem 1.1.2. To prove necessity, suppose (1.1.7) fails, that is (1.1.5) holds. By Theorem 4.3.1 and the *first part* of Theorem 4.3.3, there exists a *positive* classical solution u of (1.1.2) with equality sign (and thus also of (1.1.6) with equality), in the domain $\Omega_R = \{x \in \mathbb{R}^n : |x| > R\}$, such that $u(x) \to 0$ as $|x| \to \infty$. This violates the compact support principle. Hence (1.1.7) is necessary. □

Proof of second part of Theorem 4.3.3. Recall that (F2)′ holds by hypothesis. Then because of (1.1.7) the compact support principle Theorem 1.1.2 is valid for equation (4.3.1). But since u is a non-negative (radial) solution of (4.3.1) with $u(x) \to 0$ as $|x| \to \infty$, it now follows that u has compact support in the domain $|x| \geq R$. □

Remarks

1. The sufficiency part of Theorem 1.1.2 is closely related to Theorem 4 of [86], by specializing the results there to the matrix $a_{ij} = A(|\boldsymbol{\xi}|)\delta_{ij} + [A'(|\boldsymbol{\xi}|)/|\boldsymbol{\xi}|]\xi_i\xi_j$ which arises by expansion of the divergence term in (1.1.6). This specialization requires, however, two assumptions which are not needed here, first that the operator A be of class $C^1(\mathbb{R}^+)$, and second, that the solutions in consideration should be of class C^2 at points of Ω where $Du \neq \boldsymbol{0}$. In the proof of Theorem 4 of [86] it is also not evident that an appropriate comparison principle can be applied without the further assumption that the nonlinearity f be non-decreasing for small $u > 0$ – that is, for the validity of Theorem 4 of [86] this additional assumption, which is exactly (F2) above, seems to be required as well. For the special case of the degenerate Laplacian, see also [30].

2. The last sentence of the proof of the sufficiency of Theorem 1.1.2 gives an a priori estimate for the support of the solution u.
3. The sufficiency of condition (1.1.7) for the Compact Support Principle can be obtained under a weaker hypothesis than (F2). In fact, it is enough to suppose only

$$f(0) = 0 \quad and \; either \quad f \equiv 0 \quad or \quad F(s) > 0 \quad for \; s \in (0, \delta),$$

this condition in fact being all that it is needed for the application of Lemma 4.4.2.
4. For the case of maximal monotone graphs f, see [30], [113].

5.3   A special case

We prove here an important special case of the principal result of Section 5.4, both for its intrinsic interest as a generalization of Theorem 1.1.1 as well as to clarify the main arguments of the proof of Theorem 5.4.1. In particular, consider the differential inequality

$$\mathrm{div}\{A(|Du|)Du\} + B(x, u, Du) \le 0 \tag{5.3.1}$$

in a domain $\Omega \subset \mathbb{R}^n$. We suppose that A satisfies (A2) and a slightly stronger condition than (A1), that is

(A1)' $A \in C^1(\mathbb{R}^+)$,

and that $B(x, z, \boldsymbol{\xi}) \in L^\infty_{\mathrm{loc}}(\Omega \times \mathbb{R}^+ \times \mathbb{R}^n)$ is subject to one or the other of the conditions (B1), (B2) below:

There exist a constant $\kappa > 0$ and nonlinearities f and g, continuous in $\mathbb{R}_0^+$, such that

(B1) $B(x, z, \boldsymbol{\xi}) \ge -\kappa \Phi(|\boldsymbol{\xi}|) - f(z),$
(B2) $B(x, z, \boldsymbol{\xi}) \le \quad \kappa \Phi(|\boldsymbol{\xi}|) - g(z),$

for $x \in \Omega$, $z \ge 0$, and all $\boldsymbol{\xi} \in \mathbb{R}^n$ with $|\boldsymbol{\xi}| \le 1$.
Moreover f and g are assumed to satisfy

(F2) $f(0) = 0$ and f is non-decreasing on some interval $(0, \delta)$, $\delta > 0$;
(G2) $g(0) = 0$ and g is non-decreasing on some interval $(0, \delta)$, $\delta > 0$.

In the following results $B(x, z, \boldsymbol{\xi})$ itself need not be explicitly non-decreasing in the variable z; this corresponds to the situation of Theorem 2.1.2 where the coefficient $c(x)$ is *not* required to satisfy a sign condition for the validity of the conclusion.

Theorem 5.3.1 (Strong maximum principle). *Let* (B1) *and* (F2) *be satisfied. For the strong maximum principle to be valid for* (5.3.1) *it is sufficient that either* $f \equiv 0$ *in* $[0, d]$, $d > 0$, *or that* (1.1.5) *holds.*

Assume (B2) *and* (G2). *For the strong maximum principle to hold for* (5.3.1) *it is necessary that either* $g \equiv 0$ *in* $[0, d]$, $d > 0$, *or that*

$$\int_{0+} \frac{ds}{H^{-1}(G(s))} = \infty, \tag{5.3.2}$$

where $G(u) = \int_0^u g(s)ds$.

Proof. Sufficiency. Assume that (5.3.2) is valid. As in the proof of Theorem 1.1.1, we apply the Hopf comparison technique. Assume, contrary to the validity of the strong maximum principle, that there is a non-negative solution $u \in C^1(\Omega)$ of (5.3.1) which vanishes at some point, but is not identically zero. As in the demonstration of the Hopf Maximum Principle, Section 2.8, there is a ball B_R, with closure in Ω, such that $u > 0$ in B_R and $u = 0$ at some point $y \in \partial B_R \cap \Omega_0$, where $\Omega_0 = \{x \in \Omega : u(x) = 0\}$. Clearly $u(y) = |Du(y)| = 0$ and R can be taken arbitrarily small so that

$$0 < u < \delta, \qquad |Du| \le 1 \quad \text{in } B_R.$$

Hence by (B1) the function u is also a solution of

$$\operatorname{div}\{A(|Du|)Du\} - \kappa\Phi(|Du|) - f(u) \le 0 \quad \text{in } E_R, \tag{5.3.3}$$

where $E_R = B_R \setminus \overline{B}_{R/2}$.

Call x_0 the center of B_R. Also let $m' > 0$ be the minimum of u on $\partial B_{R/2}$ and choose

$$k = n + \kappa R.$$

As comparison function we take the non-negative radial solution $v : E_R \to \mathbb{R}^+$ of (4.2.1) given by Lemma 4.2.3, *in the space dimension k rather than n,* that is v as a function of r, $r = |x - x_0|$, satisfies the ordinary differential equation

$$[r^{k-1}\Phi(|v'|)]' + r^{k-1}f(v) = 0, \qquad v \ge 0$$

in $(R/2, R)$. For later purposes one can take the corresponding boundary value parameter m so small that $m \le m'$. In turn, in contrast with (5.3.3),

v becomes a solution of the inequality:

$$\begin{aligned}
\operatorname{div}\{&A(|Dv|)Dv\} - \kappa\Phi(|Dv|) - f(v) \\
&= -r^{1-n}\{[r^{n-1}\Phi(|v'|)]'\} - \kappa\Phi(|v'|) - f(v) \\
&= -r^{1-k}\{[r^{k-1}\Phi(|v'|)]'\} + \kappa(R/r - 1)\Phi(|v'|) - f(v) \\
&\geq -r^{1-k}\{[r^{k-1}\Phi(|v'|)]'\} - f(v) = 0
\end{aligned} \tag{5.3.4}$$

in E_R with, see Lemma 4.2.3,

$$v = 0 \text{ on } \partial B_R, \quad v = m \text{ on } \partial B_{R/2}; \quad \partial_\nu v < 0 \text{ on } \partial B_R, \quad |Dv| > 0 \text{ in } \overline{E}_R.$$

We can now apply Theorem 3.6.5 to the solutions u of (5.3.3) and v of (5.3.4) in the set E_R – that is, with the set Ω replaced by E_R. In making this application one must of course check that the principal hypotheses (i)–(ii), see Section 3.5, are verified for $A(\boldsymbol{\xi}) = A(|\boldsymbol{\xi}|)\boldsymbol{\xi}$, with $A(\mathbf{0}) = \mathbf{0}$ and with the regular set $P = \mathbb{R}^n \setminus \{\mathbf{0}\}$. This, however, follows directly from (A1)' and (A2).

To verify the further assumptions of Theorem 3.6.5, we see by (F2) that the function $-\kappa\Phi(|\boldsymbol{\xi}|) - f(z)$ is non-increasing in the variable z in the solution range $[0, \delta)$, while by (A1)' it is locally Lipschitz continuous when $\boldsymbol{\xi}$ is in P. Finally, since $Dv > 0$ it is evident that $(E_R)_v \equiv \{x \in E_R : Dv(x) \in P\} = E_R$.

Because $u \geq v$ on ∂E_R, we then obtain from Theorem 3.6.5 that $u \geq v$ in E_R. In particular $0 = \partial_\nu u(y) \leq \partial_\nu v(y) < 0$, which is a contradiction. The sufficiency part of the theorem is therefore proved. $\qquad\square$

Necessity. For each $x_0 \in \Omega$ we shall exhibit a subdomain Ω', with $x_0 \in \Omega'$, and a solution u of (5.3.1) in Ω' such that $u(x_0) = 0$ but $u \not\equiv 0$ in Ω'. The assumption to be made for this purpose is that (B2) and (G2) hold, with $g(z) > 0$ for $z > 0$, together with the negation of (5.3.2), namely

$$\int_{0+} \frac{ds}{H^{-1}(G(s))} < \infty. \tag{5.3.5}$$

Thus fix $x_0 \in \Omega$ and let $B_R \subset \Omega$ be a ball centered at x_0 with radius R. Define

$$\sigma = (n + \kappa R)^{-1},$$

where κ is given by (B2). Let C_σ be given by (4.4.1), with F replaced by G. Then choose $C < \min\{R, C_\sigma\}$, also so small that $H^{-1}(G(\gamma)) \leq 1$, where the parameter $\gamma = \gamma(C) > 0$ is defined in Lemma 4.4.1.

Put $\varepsilon = R - C$ and consider the function w given by Lemma 4.4.1 corresponding to the given value σ. For $x \in B_R$ we define the radial function $u(r) = w(r - \varepsilon)$ when $r \in [\varepsilon, R]$, $r = |x - x_0|$, and extend u as a non-negative C^1 function to all of B_R by putting $u \equiv 0$ for $0 \leq r < \varepsilon$. Then $|Du| = u' \leq 1$ in B_R by (ii) of Lemma 4.4.1, and so by (B2),

$$
\begin{aligned}
\operatorname{div}&\{A(|Du|)Du\} + B(x, u(x), Du(x)) \\
&\leq \operatorname{div}\{A(|Du|)Du\} + \kappa\Phi(u') - g(u) \\
&\leq [\Phi(u')]' + \left(\kappa + \frac{n-1}{r}\right)\Phi(u') - g(u) \\
&\leq \sigma g(u) + \left(\kappa + \frac{n-1}{r}\right)(r - \varepsilon)\,\sigma g(u) - g(u) \\
&\leq [\sigma(n + \kappa R) - 1]g(u) = 0;
\end{aligned}
$$

here we use (iii) and (iv) of Lemma 4.4.1.

But u vanishes in $B_\varepsilon(x_0)$, while $u(x) = w(C) = \gamma > 0$ when $|x| = R$, that is $u \not\equiv 0$ in $\Omega' = B_R$, contradicting the validity of the strong maximum principle. □

5.4 Strong maximum principle: Generalized version

Consider the differential inequality

$$
\partial_{x_j}\{a_{ij}(x, u)A(|Du|)\partial_{x_j}u\} + B(x, u, Du) \leq 0 \tag{5.4.1}
$$

in a domain $\Omega \subset \mathbb{R}^n$, where the symmetric coefficient matrix $\boldsymbol{a}(x, z) = [a_{ij}(x, z)]$, $i, j = 1, \dots, n$, is defined and continuously differentiable in $\Omega \times [0, \delta']$ for some $\delta' > 0$, and furthermore is such that

$$
\lambda(z)|\boldsymbol{\zeta}|^2 \leq a_{ij}(x, z)\zeta_i\zeta_j \leq \Lambda(z)|\boldsymbol{\zeta}|^2 \quad \text{for all } \boldsymbol{\zeta} \in \mathbb{R}^n, \tag{5.4.2}
$$

where λ and Λ are positive and continuous in $[0, \delta']$. We suppose that $A = A(s)$ satisfies of (A1)$'$ and (A2) of Section 5.3. Moreover $B(x, z, \boldsymbol{\xi}) \in L^\infty_{\mathrm{loc}}(\Omega \times \mathbb{R}^+ \times \mathbb{R}^n)$ is subject to one or the other of the conditions (B1) or (B2), while f and g verify (F2) and (G2). As in Section 5.3 the function $B(x, z, \boldsymbol{\xi})$ need not be explicitly non-decreasing in the variable z.

For simplicity, in the sequel we can assume without loss of generality that $\delta \leq \delta'$.

Theorem 5.4.1 (Strong maximum principle). *Suppose that*

$$\lim_{s\downarrow 0} \frac{sA'(s)}{A(s)} = c > -1 \qquad (5.4.3)$$

and, when $c \neq 0$, assume also that the positive definite matrix $[a_{ij}]$ satisfies (5.4.2) and

$$\sqrt{\frac{\Lambda(0)}{\lambda(0)}} < \phi(c), \qquad \phi(c) = \frac{2 + c + 2\sqrt{1+c}}{|c|}. \qquad (5.4.4)$$

Let (B1) and (F2) be satisfied. For the strong maximum principle [1] to be valid for (5.4.1) it is sufficient that either $f \equiv 0$ in $[0, d]$, $d > 0$, or that (1.1.5) holds.

Assume (B2) and (G2). For the strong maximum principle to hold for (5.4.1) it is necessary that either $g \equiv 0$ in $[0, d]$, $d > 0$, or that (5.3.2) is satisfied.

Proof. Sufficiency. As in the proof of Theorem 5.3.1, we apply the Hopf comparison technique. Assume, contrary to the validity of the strong maximum principle, that there is a non-negative solution $u \in C^1(\Omega)$ of (5.4.1) which vanishes at some point, but is not identically zero. As in the proof of the Hopf Maximum Principle, Section 4.5.6, there is a ball B_{R_0}, with $R_0 \leq 1$ and closure in Ω, such that $u > 0$ in B_{R_0} and $u = 0$ at some point $y \in \partial B_{R_0} \cap \Omega_0$, where $\Omega_0 = \{x \in \Omega : u(x) = 0\}$. Clearly $u(y) = |Du(y)| = 0$.

By B_R we denote any ball of radius $R \leq R_0$ which is tangent at y to ∂B_R. To begin with we take $R_1 \leq R_0$ so small that $u < \delta$ and $|Du| \leq 1$ in B_R, $R \leq R_1$. Hence, for any fixed $R \leq R_1$, by (B1) we have

$$\partial_{x_i}\{a_{ij}(x, u)A(|Du|)\partial_{x_j}u\} - \kappa\Phi(|Du|) - f(u) \leq 0 \qquad \text{in } B_R. \qquad (5.4.5)$$

We now construct an appropriate comparison function v. Define

$$\hat{a}(x) = [\hat{a}_{ij}(x)] \equiv [a_{ij}(x, u(x))],$$

obviously continuously differentiable in Ω. Define E_R to be the annular region $B_R \setminus \overline{B}_{R/2}$. Let α be a constant such that

$$|\partial_{x_i}\hat{a}_{ij}(x)| \leq \alpha$$

---

[1] The definition of the strong maximum principle is given in the first paragraph before Theorem 1.1.1.

for all $x \in E_R$. Clearly such a constant exists since $|Du| \leq 1$ and E_R is a pre-compact subset of Ω. Letting x_0 denote the center of B_R, we define $z = (x - x_0)/r$, $r = |x - x_0|$. It is then easy to see that in E_R,

$$\partial_{x_i}(\hat{a}_{ij}(x)z_j) = [\partial_{x_i}\hat{a}_{ij}(x)]z_j + \frac{\hat{a}_{ij}(x)}{r}(\delta_{ij} - z_iz_j).$$

Without loss of generality we assume that $\lambda(0) \leq 1$. Introduce

$$\lambda = \min\{\lambda(u(x)) : x \in \overline{B_R}\}, \qquad \Lambda = \max\{\Lambda(u(x)) : x \in \overline{B_R}\}.$$

Hence from (5.4.2), with $\zeta = z$,

$$|\partial_{x_i}(\hat{a}_{ij}(x)z_j)| \leq \alpha + \frac{n-1}{r}\Lambda \quad \text{for all } x \in E_R. \tag{5.4.6}$$

Put

$$k = \frac{n\Lambda + \alpha + \kappa}{\lambda} > 1;$$

of course $k = k(R)$ is uniformly bounded for all $R \leq R_1$.

   Also let $\min_{\partial B_{R/2}} u = m' > 0$ and choose $m \leq m' < \delta$. Of course m' itself depends on R.

   As comparison function v we take the radial solution $v = v(r)$, $r = |x - x_0|$, given by Lemma 4.2.3 *in the space dimension* k rather than n and with f replaced by f/λ. That is, v satisfies

$$[r^{k-1}\Phi(|v'|)]' + r^{k-1}f(v)/\lambda = 0, \qquad v \geq 0$$

in $(R/2, R)$. Now we can carry out the principal calculation, with $z = (x - x_0)/r$,

$$\partial_{x_i}\{\hat{a}_{ij}(x)A(|Dv|)\partial_{x_j}v\} - \kappa\Phi(|Dv|) - f(v)$$
$$= -\hat{a}_{ij}(x)z_iz_j[\Phi(|v'|)]' - \partial_{x_i}\{\hat{a}_{ij}(x)z_j\}\Phi(|v'|) - \kappa\Phi(|v'|) - f(v)$$
$$\geq -\hat{a}_{ij}(x)z_iz_j\left\{r^{1-k}[r^{k-1}\Phi(|v'|)]' + \frac{f(v)}{\lambda}\right\} = 0$$

in E_R for all $R \leq R_1$.

   Clearly $\int_{0+} ds/H^{-1}(F(s)/\lambda) = \infty$ by Lemma 4.1.2 and (1.1.5). Therefore the final part of Lemma 4.2.3 can be applied to the comparison function v. In summary, v is a non-negative solution of

$$\partial_{x_i}\{\hat{a}_{ij}(x)A(|Dv|)\partial_{x_j}v\} - \kappa\Phi(|Dv|) - f(v) \geq 0 \quad \text{in } E_R, \quad R \leq R_2, \tag{5.4.7}$$

such that

$$v(R) = 0, \quad v(R/2) = m; \quad \partial_\nu v < 0 \quad \text{on } \partial B_R, \quad v' < 0 \quad \text{in } \overline{E}_R. \quad (5.4.8)$$

We shall apply Theorem 3.6.5 to the solutions u and v of (5.4.5) and (5.4.7) in E_R. In making this application it is convenient to write these inequalities in the form

$$\operatorname{div}\hat{\boldsymbol{A}}(x, Du) + \hat{B}(u, Du) \geq 0, \qquad \operatorname{div}\hat{\boldsymbol{A}}(x, Dv) + \hat{B}(v, Dv) \leq 0, \quad (5.4.9)$$

where $\hat{\boldsymbol{A}} = \hat{\boldsymbol{A}}(x, \boldsymbol{\xi})$ is the vector function $A(|\boldsymbol{\xi}|)\hat{a}(x)\boldsymbol{\xi}$ and $\hat{B}(z, \boldsymbol{\xi}) = -\kappa\Phi(|\boldsymbol{\xi}|) - f(z)$. To begin with we verify the ellipticity of $\hat{\boldsymbol{A}}(x, \boldsymbol{\xi})$ in $E_R \times \boldsymbol{P}$, with

$$\boldsymbol{P} = \{\boldsymbol{\xi} : 0 < |\boldsymbol{\xi}| < \tau\}$$

and τ remaining to be determined.

To this end we observe that by virtue of (5.4.4) there exists $\mu = \mu(c) > 1$ such that

$$\sqrt{\Lambda(0)/\lambda(0)} < \mu < \phi(c).$$

Therefore, since u is continuous in Ω, there exists $R_2 \leq R_1$ so small that

$$\sqrt{\Lambda/\lambda} < \mu \qquad \text{in } B_R \qquad\qquad (5.4.10)$$

for all $R \leq R_2$.

Let d_1, d_2 be defined by $\phi(d_1) = \phi(d_2) = \mu$, $d_1 < 0 < d_2$ (see Figure 1). By (5.4.3) there exists $\tau = \tau(c) \in (0, 1]$ so small that

$$c_1 = \inf_{0 < s < \tau} \frac{sA'(s)}{A(s)} \in (d_1, c],$$

$$c_2 = \sup_{0 < s < \tau} \frac{sA'(s)}{A(s)} \in [c, d_2).$$

In turn,

$$\min\{\phi(c_1), \phi(c_2)\} > \mu. \qquad\qquad (5.4.11)$$

This being shown, let τ be the number just determined and $R \leq R_2$. Then by (5.4.10) and (5.4.11) the condition (2.4.7) in Proposition 2.4.4 is verified, with c_1, c_2 as above. Hence the matrix $[\partial_{\boldsymbol{\xi}}\hat{\boldsymbol{A}}(x, \boldsymbol{\xi})]$ is positive definite in $E_R \times \boldsymbol{P}$; that is $\hat{\boldsymbol{A}}$ is elliptic in $E_R \times \boldsymbol{P}$, as required.

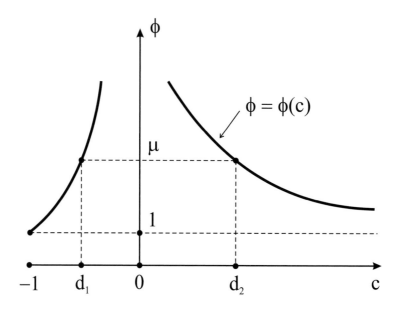

Figure 5.1: Determination of d_1 and d_2.

By (4.2.4) in Proposition 4.2.1 we can take the value m in (5.4.8) even smaller, if necessary, so that

$$0 < |Dv| < \tau \qquad \text{in } E_R, \tag{5.4.12}$$

see (5.4.8) for the first inequality.

For the application of Theorem 3.6.5 it is next necessary to check that the principal hypotheses (i)–(ii) in Section 3.5 are verified for $\hat{A}$ and $\hat{B}$, with $\boldsymbol{P} = \{\boldsymbol{\xi} : 0 < |\boldsymbol{\xi}| < \tau\}$. But these follow directly from (A1)$'$ and (A2).

From (5.4.12) moreover we see that u and v are solutions of (5.4.9) respectively in the sets $(E_R)_u$, $(E_R)_v$ (recall $\tau \leq 1$).

It remains to verify the further assumptions of Theorem 3.6.5. In particular, by (F2) the function $-\kappa\Phi(|\boldsymbol{\xi}|) - f(z)$ is non-increasing in the variable z in the solution range $[0, \delta)$, while by (A1)$'$ it is locally Lipschitz continuous when $\boldsymbol{\xi}$ is in $\boldsymbol{P}$. Finally, by (5.4.12) it is evident that $(E_R)_v = E_R$.

Because $u \geq v$ on ∂E_R, it follows from Theorem 3.6.5 that $u \geq v$ in E_R. In particular $0 = \partial_\nu u(y) \leq \partial_\nu v(y) < 0$, which is a contradiction. The sufficiency part of the theorem is therefore proved.

Necessity. For each $x_0 \in \Omega$ we shall exhibit a subdomain Ω', with $x_0 \in \Omega'$, and a solution u of (5.4.1) in Ω' such that $u(x_0) = 0$ but $u \not\equiv 0$ in Ω'. The assumption to be made for this purpose is that (B2) and (G2) hold, with $g(z) > 0$ for $z > 0$, together with the negation of (5.3.2), namely (5.3.5) holds.

Thus fix $x_0 \in \Omega$ and let B_R be a ball centered at x_0 with closure in Ω. Put $\Lambda = \max_{z \in [0,\delta]} \Lambda(z)$, and let $\alpha > 0$ be such that

$$|\partial_{x_i} a_{ij}(x, u(x))| \leq \alpha \qquad (5.4.13)$$

when $x \in B_R$, $0 \leq u(x) \leq \delta$ and $|Du(x)| \leq 1$. Clearly such a value α can be found in view of the given differentiability of $[a_{ij}]$. In turn (5.4.6) holds in B_R. Define

$$\sigma = [n\Lambda + (\alpha + \kappa)R]^{-1},$$

where κ is given by (B2).

Let C_σ be given by (4.4.1), with F replaced by G. Then choose $C < \min\{R, C_\sigma\}$, also so small that $H^{-1}(G(\gamma)) \leq 1$, where the parameter $\gamma = \gamma(C) > 0$ is defined in Lemma 4.4.1. Put $\varepsilon = R - C$ and consider the function w given by Lemma 4.4.1 corresponding to the given value σ. For $x \in B_R$ we define the radial function $u(r) = w(r - \varepsilon)$ when $r \in [\varepsilon, R]$, $r = |x - x_0|$, and extend u as a non-negative C^1 function to all of B_R by putting $u \equiv 0$ for $0 \leq r < \varepsilon$. Then $|Du| = u' \leq 1$ in B_R by (ii) of Lemma 4.4.1.

We now carry out the main calculation, with $z = (x - x_0)/r$,

$$\partial_{x_i}\{a_{ij}(x, u(x))A(|Du|)\partial_{x_j}u\} + B(x, u(x), Du(x))$$

$$\leq \partial_{x_i}\{a_{ij}(x, u(x))A(|Du|)\partial_{x_j}u\} + \kappa\Phi(u') - g(u) \qquad \text{by (B2)}$$

$$\leq a_{ij}(x, u(x))z_i z_j[\Phi(u')]' + \left(\alpha + \kappa + \Lambda\frac{n-1}{r}\right)\Phi(u') - g(u) \qquad (5.4.14)$$

$$\leq \Lambda\sigma g(u) + \left(\alpha + \kappa + \Lambda\frac{n-1}{r}\right)(r - \varepsilon)\sigma g(u) - g(u)$$

$$\leq \{\sigma[n\Lambda + (\alpha + \kappa)R] - 1\}g(u) = 0;$$

in obtaining (5.4.14) we use (5.4.6), together with (iii) and (iv) of Lemma 4.4.1.

But u vanishes in $B_\varepsilon(x_0)$, while $u(x) = w(C) = \gamma > 0$ when $|x - x_0| = R$, that is $u \not\equiv 0$ in $\Omega' = B_R$, contradicting the validity of the strong maximum principle. $\qquad\square$

It is exactly in the application of Proposition 3.6.5 at the end of the proof of sufficiency that the strengthened condition (A1)$'$ is needed.

There is also a maximum principle for the converse differential inequality

$$\partial_{x_j}\{a_{ij}(x,u)A(|Du|)\partial_{x_j}u\} + B(x,u,Du) \geq 0, \qquad u \geq 0, \quad (5.4.15)$$

in $\Omega \subset \mathbb{R}^n$, which can be obtained as an immediate consequence of Theorem 5.4.1.

Theorem 5.4.2 (Strong Maximum Principle). *Suppose that* (5.4.3) *holds, and that* (B2) *applies with* $g(z) \geq 0$ *for* $z \in [0, \delta)$. *Let* $\overline{\delta} \leq \delta$ *be such that*

$$\sup_{z \in [0,\overline{\delta})} \sqrt{\Lambda(z)/\lambda(z)} < \phi(c) \qquad (5.4.16)$$

(*when* $c = 0$ *we can take* $\overline{\delta} = \delta$).

If u *is a non-negative solution of* (5.4.15) *in* Ω, *then* u *cannot attain a maximum value* $M \in [0, \overline{\delta})$ *in the interior of* Ω, *unless* $u \equiv M$.

Proof. Suppose u reaches a maximum value M in $[0, \overline{\delta})$ in Ω. Define $\overline{u}(x) = M - u(x)$. Then $\overline{u}$ is non-negative and obeys the inequality

$$\partial_{x_j}\{a_{ij}(x, M - \overline{u})A(|D\overline{u}|)\partial_{x_j}\overline{u}\} - \kappa\Phi(|D\overline{u}|) \leq 0$$

at all points of Ω where $|D\overline{u}| < 1$. This has exactly the form (5.4.1) with

$$B(x, z, \boldsymbol{\xi}) = -\kappa\Phi(|\boldsymbol{\xi}|).$$

That is, (B1), (F2) hold with $f(z) \equiv 0$.

We can therefore apply Theorem 5.4.1 to the (non-negative) solution $\overline{u}$, provided (5.4.4) applies with $\Lambda(0)$, $\lambda(0)$ replaced by $\Lambda(M)$, $\lambda(M)$. But this is a consequence of (5.4.16), as required. Hence the strong maximum principle Theorem 5.4.1 applies to $\overline{u}$, and we get $\overline{u} \equiv 0$ in Ω, i.e., $u \equiv M$ in Ω. $\qquad\square$

Corollary 5.4.3 (Strong Maximum Principle). *Suppose that* (5.4.3) *holds, and that* (B2) *applies with* $g(z) \geq 0$ *for* $z \in [0, \delta)$. *Suppose the matrix* $[\boldsymbol{a}]$ *is independent of* u *and*

$$\sqrt{\Lambda/\lambda} < \phi(c).$$

If u *is a non-negative solution of* (5.4.15) *in* Ω, *then* u *cannot attain a maximum value* $M \in [0, \delta)$ *in the interior of* Ω, *unless* $u \equiv M$.

This result is closely related to Theorem 3.7.4 in the case $b = 0$. Theorem 5.4.1 implies as well a necessary and sufficient criterium for the validity of the strong maximum principle.

Corollary 5.4.4. *Assume* (B1), (B2), (F2), (G2), (5.4.3) *and, when* $c \neq 0$, *also* (5.4.4). *Suppose that there exists* $\nu \in (0,1]$ *such that* $g(z) \geq \nu f(z) > 0$ *for* $z \in (0, \delta)$. *Then the strong maximum principle is valid for* (5.4.1) *if and only if either* $f \equiv 0$ *in* $[0, d]$, $d > 0$, *or* (1.1.5) *holds.*

Remarks

1. When $c \neq 0$ in (5.4.3) and

$$\sqrt{\frac{\Lambda(0)}{\lambda(0)}} > \frac{2 + c + 2\sqrt{1 + c}}{|c|} \tag{5.4.17}$$

   in Theorem 5.4.1, the proof given above fails, since by Theorem 1.3 of [20] the matrix $[\partial_{\xi_j} \hat{A}_i(x, \boldsymbol{\xi})]$ can be indefinite for some directions ζ of the vector Du and for some points $x \in \Omega$. Of course exactly such points and directions occur when the normal at the tangent point $x = x_0$ is a direction $\boldsymbol{\nu}$. Thus the proof of Theorem 5.4.1 fails in this case, since x_0 could be any point in Ω and the normal could have *any* direction $\boldsymbol{\nu}$, depending on the particular outcome of the Hopf construction.

2. It is *an open question* whether Theorem 5.4.1 itself fails when (5.4.17) is valid. We have not been able to find a counterexample for such cases, though it may be conjectured that the condition (5.4.4) is in fact necessary.

3. Condition (5.4.4) is automatically valid if $a(x, 0)$ is a positive multiple of the identity. Indeed, if $a_{ij}(x, u) = a(x, u)\delta_{ij}$, where $a : \Omega \times \mathbb{R}_0^+ \to \mathbb{R}^+$ is of class C^1, then the differential operator in (5.4.1) has the form

$$\text{div}\{a(x, u)A(|Du|)Du\}.$$

   For this special case, Theorem 5.4.1 continues to hold without the help of (5.4.4), since $\Lambda(0)/\lambda(0) = 1$ and so $\partial_{\boldsymbol{\xi}} \hat{A}(x, \boldsymbol{\xi})$ is positive definite without further argument.

4. Condition (5.4.3) applies to the p-Laplace operator $A(s) = s^{p-2}$, $p > 1$, with $c = p - 2$. In this case, when $c \neq 0$, namely when $p \neq 2$, the condition (5.4.4) takes the explicit form

$$\sqrt{\frac{\Lambda(0)}{\lambda(0)}} < \frac{p + \sqrt{p - 1}}{|p - 2|}. \tag{5.4.18}$$

5. The validity of Theorem 5.4.1 can obviously be asserted if the differential inequality (5.4.1) is *assumed* to be elliptic for all arguments $(x, u, Du) \in \Omega \times \mathbb{R}_0^+ \times \mathbb{R}^n$ such that $0 < u < \varepsilon$, $0 < |Du| < \varepsilon$ for some $\varepsilon > 0$.

If $c = 0$ in (5.4.3), as occurs for example when $A(s) = 1$, i.e., for the Laplace operator, or when $A(s) = 1/\sqrt{1 + s^2}$, i.e., the mean curvature operator, then condition (5.4.4) is empty and so Theorem 5.4.1 is correct even with no additional conditions on $[a_{ij}]$ outside of positive definiteness and regularity! This yields

Theorem 5.4.5. *Assume* (B1), (F2). *Then the strong maximum principle is valid for the mean curvature type differential inequality*

$$\partial_{x_i} \left\{ \frac{a_{ij}(x, u)\partial_{x_j} u}{\sqrt{1 + |Du|^2}} \right\} + B(x, u, Du) \leq 0, \qquad u \geq 0 \qquad in \ \Omega \quad (5.4.19)$$

if either $f \equiv 0$ in $[0, d]$, $d > 0$, or (1.1.5) is satisfied.

Assume (B2), (G2). *For the strong maximum principle to hold for* (5.4.19) *it is necessary that either $g \equiv 0$ for $u \in [0, d]$, $d > 0$, or that* (5.3.2) *holds.*

Here it is worth noting that (5.4.19) is *not* elliptic exactly at points where

$$|Du| > \frac{2 \sqrt[4]{\ell}}{\sqrt{\ell} - 1}, \qquad \ell = \frac{\Lambda(u)}{\lambda(u)}.$$

Example: the linear case. Consider the linear inequality

$$\partial_{x_i}\{a_{ij}(x)\partial_{x_j} u\} + b_i(x)\partial_{x_i} u + c(x)u \leq 0, \qquad u \geq 0, \qquad (5.4.20)$$

for $x \in \Omega$, where the matrix $[a_{ij}]$ is continuously differentiable and satisfies (5.4.2) with λ, Λ, independent of z while b_i, $c \in C(\Omega)$ for all $i = 1, \ldots, n$. This is the special case of (5.4.1) where $A(s) \equiv 1$, $B(x, z, \boldsymbol{\xi}) = b_i(x)\xi_i + c(x)z$. Here we can apply the result of Theorem 5.4.1, assuming also that $b_i(x)$ and $c(x)$ are locally bounded. By slightly shrinking the domain Ω we can then suppose that

$$\kappa = \max_i \sup_\Omega |b_i(x)| < \infty, \qquad d = -\inf_\Omega \{c(x), 0\} < \infty.$$

Moreover define $f(z) = dz$. Then $\Phi(s) = s$, $H^{-1}(s) = \sqrt{2s}$ and $F(z) = dz^2/2$, so that (B1) and (1.1.5) hold as required; here $c = 0$ in (5.4.3).

This gives the strong maximum principle for (5.4.20), closely related to the classical Theorem 2.1.2 of E. Hopf. Indeed, the strong maximum principle for C^2 solutions of (5.4.20) is an immediate consequence of Theorem 2.1.2, while conversely the strong maximum principle for C^1 weak solutions of (2.1.1), written in the form (5.4.20), follows at once from Theorem 5.4.1.

These comments moreover lead us to expect that the proof of Theorem 5.4.1 can be simplified for the special linear case. In fact, the proof of Theorem 5.4.1 suggests that the required comparison function v can be obtained for the linear case by exhibiting an explicit solution of the inequality

$$v'' + \frac{k}{r}v' - \frac{dv}{\lambda} \geq 0, \qquad v' \leq 0$$

(since $\Phi(s) = s$ in the present linear case). A natural choice for v is

$$v(r) = \alpha[(R/r)^\vartheta - 1], \qquad R/2 \leq r \leq R, \qquad (5.4.21)$$

where ϑ and R are to be determined. Then $v'(r) \leq 0$ and a short calculation gives

$$v'' + \frac{k}{r}v' - \frac{dv}{\lambda} = \alpha\vartheta \left(\frac{R}{r}\right)^\vartheta \left\{\frac{(\vartheta+1)-k}{r^2}\right\} - \frac{dv}{\lambda} \geq \alpha \left\{\frac{(\vartheta+1)-k}{R^2} - \frac{d}{\lambda\vartheta}\right\}.$$

This will be ≥ 0 provided that

$$\vartheta = 2k - 1, \qquad R^2 \leq \frac{\lambda k(2k-1)}{d}.$$

Thus the rational comparison function (5.4.21) can be used for the linear inequality (5.4.20), alternative to the standard exponential function

$$v(r) = \varepsilon\left(e^{-\alpha r^2} - e^{-\alpha R^2}\right),$$

see page 148 of [46], or page 34 of [43].

5.5 A boundary point lemma

Equation (5.4.1) also has a corresponding boundary point lemma. Remarkably, in contrast with Hopf's boundary point lemma, the basic equation need not be uniformly elliptic, this ultimately being due to the strong result of Lemma 4.2.3.

Theorem 5.5.1 (Boundary Point Lemma). *Assume* (5.4.3) *and when* $c \neq 0$ *also* (5.4.4). *Suppose that* (B1), (F2) *hold and that either* $f \equiv 0$ *in* $[0, d]$, $d > 0$, *or that* (1.1.5) *is satisfied.*

Let u *be a* C^1 *solution of* (5.4.1) *in* $\overline{\Omega}$, *with* $u > 0$ *in* Ω *and* $u(y) = 0$, *where* $y \in \partial\Omega$. *If* Ω *satisfies an interior sphere condition at* y, *then* $\partial_\nu u < 0$ *at* y.

Proof. By the interior sphere condition there exists an open ball $B_R = B_R(x_0) \subset \Omega$ with $y \in \partial B_R$. If R is suitably small, then there exists, exactly as in the proof of the sufficiency of Theorem 5.4.1, a comparison function v in the annular region $E_R = B_R \setminus \overline{B}_{r/2}$. Continuing as in the proof of Theorem 5.4.1 it follows that $u \geq v$ in E_R, which immediately supplies the conclusion $\partial_\nu u(y) \leq \partial_\nu v(y) = v'(R) < 0$. □

There is also a boundary point lemma corresponding to Theorem 5.4.2.

Theorem 5.5.2. *Assume that the hypotheses of Theorem 5.4.2 are satisfied. Let* u *be a* C^1 *solution of* (5.4.15) *in* $\overline{\Omega}$, *with* $0 \leq u < M$ *in* Ω *and* $u(y) = M$, *where* $y \in \partial\Omega$. *If* $M \in [0, \overline{\delta})$ *and* Ω *satisfies an interior sphere condition at* y, *then* $\partial_\nu u > 0$ *at* y.

The proof is essentially the same as for Theorem 5.5.1, but using the transformation $\overline{u} = M - u$ as in Theorem 5.4.2.

5.6 Compact support principle: Generalized version

Here we consider the converse inequality

$$\partial_{x_i}\{a_{ij}(x)A(|Du|)\partial_{x_j}u\} + B(x, u, Du) \geq 0, \tag{5.6.1}$$

the domain Ω being an exterior set, say with $\Omega \supset \Omega_R = \{x \in \Omega : |x| > R\}$. Conditions (A1)′, (A2) and (5.4.2) are assumed to be valid, as for the Strong Maximum Principle, along with one or the other of conditions (B1), (B2), (F1), (F2) of Section 5.3.

Here we restrict the matrix $[a_{ij}]$ to depend only on x, with the coefficients $a_{ij}(x)$ having uniformly bounded derivatives in Ω. The functions λ and Λ in (5.4.2) are now purely positive constants. (A corresponding boundedness condition on the derivatives of a_{ij} is unneeded for the Strong Maximum Principle because the arguments there are purely local. Note also that the functions λ, Λ in (5.4.2) are now simply positive constants.)

Theorem 5.6.1 (Compact Support Principle[2]). *If* (B1) *and* (F2) *are sat-isfied, with* $f(z) > 0$ *for* $z > 0$, *then for the compact support principle to hold for* (5.6.1) *it is necessary that* (1.1.7) *be valid.*

On the other hand, assume (5.4.3), *and when* $c \neq 0$ *that*

$$\sqrt{\Lambda/\lambda} < \phi(c).$$

Then for the compact support principle to hold for (5.6.1) *it is sufficient that* (B2) *and* (G2) *are satisfied, with* $g(z) > 0$ *for* $z > 0$ *and*

$$\int_{0+} \frac{ds}{H^{-1}(G(s))} < \infty. \tag{5.6.2}$$

Proof. Necessity. Here it will be enough to show the existence of a (radial) solution $u = u(r)$ of the following problem in the exterior domain Ω_R,

$$\begin{cases} \partial_{x_i}\{a_{ij}(x)A(|Du|)\partial_{x_j}u\} + B(x, u, Du) \geq 0, & \text{in } \Omega_R, \\ u(R) = m, \quad u(r) \to 0 \quad \text{as } r \to \infty; \quad u > 0, \quad u' < 0 \quad \text{in } \Omega_R, \end{cases} \tag{5.6.3}$$

where (B1) and (F2) hold, with $f(z) > 0$ for $z > 0$, and also, by negation, condition (1.1.5) is satisfied.

To this end, as in the proof of sufficiency for the Strong Maximum Principle, it is enough to consider the equation

$$[\Phi(u')]' + \frac{1}{\lambda}\left(\alpha + \kappa + \frac{n-1}{r}\Lambda\right)\Phi(u') - \frac{f(u)}{\lambda} = 0,$$
$$0 < u < \delta, \quad -1 \leq u' < 0,$$

where $\alpha = \sup_{x \in \Omega}|\partial_{x_i}a_{ij}(x)|$. That is, the problem becomes

$$\begin{cases} [\tilde{q}(r)\Phi(u')]' - \tilde{q}(r)\tilde{f}(u) = 0, & \text{in } [R, \infty), \\ u(R) = m, \quad u(r) \to 0 \quad \text{as } r \to \infty, \\ u > 0, \quad -1 \leq u' < 0 \quad \text{in } \Omega_R, \end{cases} \tag{5.6.4}$$

where $' = d/dr$ and $\tilde{q}, \tilde{f}$ are given by

$$\tilde{q}(r) = r^{(n-1)\lambda^{-1}\Lambda}e^{(\alpha+\kappa)\lambda^{-1}r}, \qquad \tilde{f}(u) = f(u)/\lambda.$$

Of course, $\tilde{f}(u)$ continues to obey (1.1.7), by Lemma 4.1.2.

---

[2]For the definition of the compact support principle, see the first paragraph before Theorem 1.1.2.

The required solution can now be constructed (for $m < \delta$) exactly as in the proof of Theorem 4.3.1, with only the change that $q(r) = r^{n-1}$ is replaced by the new function $\tilde{q}(r)$, and $f(u)$ by $\tilde{f}(u)$. In particular, we can guarantee $|u'| \leq 1$ by using (4.2.4) and taking m suitably small. Moreover, one gets $u(r) \to \ell = 0$ as $r \to \infty$ by the argument at the end of the proof of Theorem 4.3.1, but with the ratio $r/(r+1)$ replaced by

$$\frac{\tilde{q}(r)}{\tilde{q}(r+1)} = e^{-(\alpha+\kappa)/\lambda} \left(\frac{r}{r+1} \right)^{(n-1)\Lambda/\lambda}.$$

This approaches the *positive* limit $e^{-(a+\kappa)\Lambda/\lambda}$ as $r \to \infty$, which gives $\ell = 0$ and completes the proof of necessity.

Sufficiency. The basic method of proof is taken from Theorem 2′ of [84], with a number of modifications.

Consider a solution $u \in C^1(\Omega)$ of the inequality (5.4.15) in an exterior domain Ω, with $u(x) \to 0$ as $|x| \to \infty$. Under the conditions (B2), (G2) it is required to show that u has compact support in Ω.

As before, put $\alpha = \sup_{x \in \Omega} |\partial_{x_i} a_{ij}(x)|$ and define

$$\sigma = (\Lambda n + \alpha + \kappa)^{-1}.$$

With the help of the End Point Lemma 4.4.1, with F replaced by G, we can now construct an appropriate radial comparison function $v = v(r)$. Let C be chosen and fixed so that

$$C < \min\{1, C_\sigma\}, \qquad H^{-1}(G(\gamma)) < \tau, \qquad \gamma = \gamma(c) < \delta,$$

where $\tau = \tau(c)$ is given in the proof of Theorem 5.4.1.

For any $R > 0$ define

$$v(r) = w(R + C - r), \qquad R \leq r \leq R + C, \quad r = |x|,$$

where w is given by (4.4.2), corresponding to the constants σ and C. By (ii) of Lemma 4.4.1,

$$v(R) = w(C) = \gamma, \qquad |Dv| < \tau. \tag{5.6.5}$$

Moreover $v'(r) < 0$ for $R \leq r < R + C$ and $v(R + C) = v'(R + C) = 0$. We can thus suppose that v is extended to all $r \geq R$ by taking $v(r) \equiv 0$ for $r > R + C$.

To check that v has the required property of an upper comparison function, we see from Lemma 4.4.1 that in the annulus $E = \{x \in \mathbb{R}^n : R < |x| < R + C\}$,

$$\partial_{x_i}\{a_{ij}(x)A(|Dv|)\partial_{x_j}v\} + \kappa\Phi(|Dv|) - g(v)$$

$$\leq a_{ij}(x)\frac{x_i x_j}{r^2}[\Phi(|v'|)]' + \left(\alpha + \kappa + \Lambda\frac{n-1}{r}\right)\Phi(|v'|) - g(v)$$

$$\leq [\sigma(\Lambda n + \alpha + \kappa) - 1]g(v) \leq 0;$$

the steps in this calculation are essentially the same as those previously used to derive (5.4.14). In summary, we have

$$\partial_{x_i}\{a_{ij}(x)A(|Dv|)\partial_{x_j}v\} + \kappa\Phi(|Dv|) - g(v) \leq 0 \qquad (5.6.6)$$

in Ω_R. Here $v \equiv 0$ for $r \geq R + C$, while $v > 0$, $-\tau < v' < 0$ for $R \leq r < R + C$.

Let R be so large that $\Omega_R \subset\subset \Omega$ and

$$0 \leq u < \gamma \, (< \delta) \qquad \text{in } \Omega_R. \qquad (5.6.7)$$

Put $M = \max_{|x|=R+C} u(x)$. Then, since $u(x) \to 0$ as $|x| \to \infty$, we see from Corollary 5.4.3 that $u \leq M$ in Ω_{R+C}. If $M = 0$ we are done. We thus assume for contradiction that $M > 0$, in which case necessarily $u < M$ in Ω_{R+C} by Corollary 5.4.3.

Let the maximum value M of $u(x)$, $|x| = R + C$, be reached at y. Then, since Ω_{R+C} obviously satisfies an interior sphere condition at y, the Boundary Point Theorem 5.5.2 applies in Ω_{R+C}. Hence $\partial_\nu u > 0$ at y.

Our purpose is now to apply the comparison Theorem 3.6.5 in the set E. To this end, we observe first, by (B2) and the fact that $\tau \leq 1$, that u is a solution of

$$\partial_{x_i}\{a_{ij}(x)A(|Du|)\partial_{x_j}u\} + \kappa\Phi(|Du|) - g(u) \geq 0, \qquad (5.6.8)$$

in $E_u = \{x \in E : |Du(x)| \in P\}$, where $P = \{\xi \in \mathbb{R}^n : 0 < |\xi| < \tau\}$. Similarly v is a solution of (5.6.6) in $E \equiv E_v$, since $Dv \in P$ when $x \in E$.

As in the proof of Theorem 5.4.1 the operator $A(\xi) = a(x)A(|\xi|)\xi$ is elliptic in $\Omega \times P$, while of course conditions (i), (ii) of Section 3.5 continue to hold.

Since by (5.6.5) and (5.6.7) we have $u \leq v + M$ on ∂E, it follows that $u \leq v + M$ in E. Thus $\partial_\nu u \leq \partial_\nu v = 0$ at y. This contradicts the previously established relation $\partial_\nu u > 0$ at y. $\square$

Corollary 5.6.2. *Assume* (B1), (B2), (F2), (G2), (5.4.3), *and when* $c \neq 0$ *also* (5.4.4). *Suppose that there exists* $\nu \in (0, 1]$ *such that* $g(z) \geq \nu f(z) > 0$ *for* $z > 0$. *Then the compact support principle is valid for* (5.6.1) *if and only if* (1.1.7) *holds.*

Remarks. It is clear from the proof of the necessity part of the Compact Support Principle Theorem 5.6.1 that the matrix $[a_{ij}]$ in this case can depend on z as well as x, since the solution u considered there, together with its gradient, is a priori bounded, see (5.6.4). That is $\alpha = \sup_{x \in \Omega_R} |(\partial_{x_i} a_{ij}(x, u(x)))_{j=1}^n|$ is still finite.

It is an open problem whether the sufficiency of the Compact Support Principle for (5.6.1) remains valid when the matrix $[a_{ij}]$ is also allowed to depend on the solution variable z.

The following counterexample [84] shows the importance of the boundedness condition (B2). Consider the inequality

$$\Delta_p u + |Du|^{q_1} - u^{q_2} \geq 0, \quad u \geq 0, \quad p > 1, \ q_1, q_2 > 0. \quad (5.6.9)$$

Clearly (5.4.3) holds with $c = p - 2$, and conditions (5.6.2) and (B2) are satisfied if and only if $q_1 \geq p - 1$ and $q_2 < p - 1$. The compact support principle then holds for (5.6.9). On the other hand, for any $q_1 \in (0, p - 1)$ we can take $q_1 < q_2 < p - 1$. One easily checks that (5.6.9) then has positive solutions $u(x) = \text{const.} |x|^{-\kappa}$ on Ω_R for κ and R suitably large. Hence the compact support principle fails even though condition (5.6.2) is fulfilled!

The case $c = 0$ in (5.4.3) can be treated exactly as in Section 5.4, leading to the following result for the mean curvature type inequality.

Theorem 5.6.3. *Assume* (B2) *and* (G2) *are satisfied, with* $g(z) > 0$ *for* $z > 0$, *and that* (5.6.2) *holds. Then the compact support principle is valid for the mean curvature type differential inequality*

$$\partial_{x_i} \left\{ \frac{a_{ij}(x) \partial_{x_j} u}{\sqrt{1 + |Du|^2}} \right\} - B(x, u, Du) \geq 0, \quad in \ \Omega. \quad (5.6.10)$$

On the other hand, if (B2) *and* (F2) *are valid, with* $f(z) > 0$ *for* $z > 0$, *then for the compact support principle to hold for* (5.6.10) *it is necessary that* (1.1.7) *be satisfied.*

Notes

The background and literature for Theorem 1.1.1 is fairly complicated and deserves a number of comments.

The necessity of (1.1.5) for the case of the Laplace operator is due to Benilan, Brezis and Crandall [10], while for the p-Laplacian it is due to Vázquez [113]. In these cases (1.1.5) reduces respectively to

$$\int_{0+} \frac{ds}{\sqrt{F(s)}} = \infty \quad \text{and} \quad \int_{0+} \frac{ds}{[F(s)]^{1/p}} = \infty.$$

For general operators satisfying (A1), (A2), necessity is due to Diaz ([28], Theorem 1.4).

Sufficiency for the case of the Laplace operator and also for the p-Laplacian is again due to Vázquez [113], see also [28], [50] and [69]. For general operators satisfying (A1), (A2), sufficiency was proved in Theorem 1 of [84] under an additional technical assumption, in Theorem 1 of [79] without the technical assumption and in Theorem 1.1 of [81] with a simplified proof. For the vectorial case see [36].

The case when $f \equiv 0$ was studied by Cellina [19] for non-negative minimizers of the integral $\int_\Omega \mathscr{G}(|Du|)dx$. An alternative abstract approach to the strong maximum principle appears in [21].

As in the case of the strong maximum principle it is worth commenting on the background and literature for the compact support principle Theorem 1.1.2.

Necessity was first shown in Corollary 2 of [84] under an additional technical assumption as noted above, and in [79], with a proof which is not at all easy. The proof given in [81] is simpler and at the same time provides an existence theorem for radial solutions of exterior Dirichlet problems; see Theorem 4.3.1.

The sufficiency of (1.1.7) is Theorem 2 of [84]. For radially symmetric solutions of (1.1.6) sufficiency was proved in Proposition 1.3.1 of [39] under the weaker assumption that $F(z) > 0$ for $z \in (0, \delta)$.

For the generalized versions of the strong maximum principle and the compact support principle, see [84] and [20]; the proofs here are shortened and improved. Theorem 5.5.1 includes the Hopf boundary point lemma, together with extensions to divergence structure inequalities drawn from [94].

Problems

5.1 Consider the divergence structure operator $\operatorname{div}(A(|Du|)Du)$, and suppose the function $A = A(s)$, $s > 0$, is positive and continuously differentiable and that $\{sA(s)\}' > 0$ for all $s > 0$ and $sA(s) \to 0$ as $s \to 0$. Show that the corresponding non-divergence structure quasilinear operator is elliptic for functions $u \in C^2(\Omega)$ with $Du \neq \mathbf{0}$. If moreover A is positive and continuously differentiable for $s \geq 0$ and $\{sA(s)\}' > 0$ for all $s \geq 0$, the corresponding operator is uniformly elliptic for any function $u \in C^2(\Omega)$ with $|Du|$ bounded in Ω.

The conditions (A1) and (A2) in the Introduction are a generalization for the operator of the standard notion of ellipticity for quasilinear operators.

5.2 Show that the operators $A(s) = s^{p-2}$, $p > 1$, $A(s) = 1/\sqrt{1+s^2}$ satisfy conditions (A1) and (A2). For what values of the exponents a and b does $A(s) = (1 + s^a)^b$ satisfy (A1) and (A2)? What are the corresponding functions $\mathscr{G}$, assuming $\mathscr{G}(0) = 0$? Find $H(s)$ when $A(s) = (1 + s^2)^b$.

5.3 Verify the conditions given in Section 1.1 for the functions $u(x) = C|x|^k$ and $v(x) = L|x|^{-\ell}$ to satisfy (1.1.10).

Chapter 6

Non-homogeneous Divergence Structure Inequalities

6.1 Maximum principles for structured inequalities

We consider the quasilinear differential inequality

$$\operatorname{div} \boldsymbol{A}(x, u, Du) + B(x, u, Du) \geq 0 \qquad \text{in } \Omega, \qquad (6.1.1)$$

where Ω is a bounded domain in $\mathbb{R}^n$, and $\boldsymbol{A}$ and B satisfy the generic assumptions of Section 3.1. Here we shall extend the validity of Theorems 3.2.1 and 3.2.2 to the case when (6.1.1) is inhomogeneous, that is, there are constants a_2, b_1, b_2, a, $b \geq 0$ such that for all $(x, z, \boldsymbol{\xi}) \in \Omega \times \mathbb{R}^+ \times \mathbb{R}^n$ there holds, for $p > 1$,

$$
\begin{aligned}
\langle \boldsymbol{A}(x, z, \boldsymbol{\xi}), \boldsymbol{\xi} \rangle &\geq |\boldsymbol{\xi}|^p - a_2 z^p - a^p, \\
B(x, z, \boldsymbol{\xi}) &\leq b_1 |\boldsymbol{\xi}|^{p-1} + b_2 z^{p-1} + b^{p-1},
\end{aligned}
\qquad (6.1.2)
$$

while for $p = 1$,

$$\langle \boldsymbol{A}(x, z, \boldsymbol{\xi}), \boldsymbol{\xi} \rangle \geq |\boldsymbol{\xi}| - a_2 z - a, \qquad B(x, z, \boldsymbol{\xi}) \leq b \qquad (6.1.3)$$

(in (6.1.3) we write b for b_2 and discard the terms $b_1|\boldsymbol{\xi}|^{p-1}$, b^{p-1}). As in Section 3.1 the domain Ω is assumed to be bounded. This condition can be removed if Ω has finite measure and the boundary condition for $|x| \to \infty$ is taken in the form (3.2.12).

The apparently more general situation when the principal term $|\boldsymbol{\xi}|^p$ in (6.1.2) is replaced by $a_1|\boldsymbol{\xi}|^p$, $a_1 > 0$, in fact immediately reduces to (6.1.2) by rescaling.

In the following results we deal with p-regular solutions, without further mention.

Theorem 6.1.1 (Semi-maximum principle). *Let $u \in W_{loc}^{1,p}(\Omega)$, $p \geq 1$, be a solution of the inequality (6.1.1) in Ω, with $u \leq M$ on $\partial\Omega$ for some constant $M \geq 0$. If (6.1.2) holds, then $u^+ \in L^\infty(\Omega)$ and*

$$u \leq C[\|u^+\|_p + k + (a_2^{1/p} + b_2^{1/(p-1)})M] + M \qquad a.e. \ in \ \Omega, \quad (6.1.4)$$

where $k = a + b \geq 0$ and the constant C depends only on p, n, $|\Omega|$, b_1 and $a_2 + b_2$.

(If $p = 1$, then $b_2^{1/(p-1)}$ is dropped from (6.1.4), $k = a$ and the constant C depends on n, $|\Omega|$ and $a_2 + b$.)

When $p < n$ an explicit form for the constant C in Theorem 6.1.1 can be obtained from (6.2.18), (6.2.28) with $\varepsilon = 1$, and (6.2.27). The same holds for $p \geq n$, except that (6.2.18) should be replaced by (6.2.24) with $\varepsilon = 1$. A similar remark applies to the following results.

Theorem 6.1.2. *Theorem 6.1.1 continues to be valid if the coefficients a, b, a_2, b_1 and b_2 are functions in the Lebesgue spaces:*

$$a, b_1 \in L^{p\alpha}(\Omega), \quad b \in L^{(p-1)\alpha}(\Omega), \quad a_2, b_2 \in L^\alpha(\Omega),$$

$$\alpha = \frac{\max\{n/p, 1\}}{1 - \varepsilon}, \quad \varepsilon \in (0, 1] \qquad (6.1.5)$$

and (6.1.4) is replaced by

$$u \leq C[\|u^+\|_p + k + (\|a_2\|^{1/p} + \|b_2\|^{1/(p-1)})M] + M, \quad a.e. \ in \ \Omega, \quad (6.1.6)$$

where $k = \|a + b\|$ and the constant C now depends also on ε.

Here and in the sequel, we understand by $\|a\|$, $\|b\|$, $\|a_2\|$, $\|b_1\|$ and $\|b_2\|$ the norms of a, b, a_2, b_1, b_2 in the respective Lebesgue spaces (6.1.5), or, in the limit case $\varepsilon = 1$, the Lebesgue space $L^\infty(\Omega)$.

(If $p = 1$ then b_1 and b_2 should be omitted from (6.1.5) while a, b, $a_2 \in L^{n/(1-\varepsilon)}(\Omega)$ and $k = \|a\|$.)

Theorem 6.1.3 (Maximum principle). *Let* $u \in W_{loc}^{1,p}(\Omega)$, $p > 1$, *be a solution of* (6.1.1) *in* Ω, *where* $\mathbf{A}$ *and* B *satisfy* (6.1.2) *with* $b_1 = b_2 = 0$. *Suppose* $u \leq M$ *on* $\partial\Omega$ *for some constant* $M \geq 0$. *Then* $u^+ \in L^\infty(\Omega)$ *and*

$$u \leq C(a + b + a_2^{1/p}M) + M \qquad a.e. \ in \ \Omega, \tag{6.1.7}$$

where C *can be taken in the form* $\exp\{C(p, n, |\Omega|)(1 + a_2)^\nu\}$ *with* $\nu = (n + p)/p^2$ *when* $1 < p < n$ *and* $\nu = 5/p$ *when* $p \geq n$.

Theorem 6.1.4 (Maximum principle). *Let* $u \in W_{loc}^{1,p}(\Omega)$, $p > 1$, *be a solution of* (6.1.1) *in* Ω, *where* $\mathbf{A}$ *and* B *satisfy* (6.1.2) *with* $a_2 = b_2 = 0$. *Suppose* $u \leq M$ *on* $\partial\Omega$ *for some constant* $M \geq 0$. *Then* $u^+ \in L^\infty(\Omega)$ *and*

$$u \leq C(a + b) + M \qquad a.e. \ in \ \Omega, \tag{6.1.8}$$

where C *can be taken in the form* $\exp\{C(p, n, |\Omega|)(1 + b_1)^{(1+n)/p}\}$.

Theorem 6.1.5. *Theorems* 6.1.3 *and* 6.1.4 *continue to be valid if the coefficients* a, b, a_2 *and* b_1 *are functions in the Lebesgue spaces:*

$$a, b_1 \in L^{p\beta}(\Omega), \quad b \in L^{(p-1)\beta}(\Omega), \quad a_2 \in L^\beta(\Omega),$$

$$\beta = \begin{cases} n/p(1 - \varepsilon), & if \ 1 < p \leq n, \\ 1, & if \quad p > n, \end{cases} \quad \varepsilon \in (0, 1]. \tag{6.1.9}$$

That is, the estimate (6.1.7) *becomes*

$$u \leq C(\|a\| + \|b\| + \|a_2\|^{1/p}M) + M \qquad a.e. \ in \ \Omega, \tag{6.1.10}$$

and similarly (6.1.8) *changes into* $u \leq C(\|a\| + \|b\|) + M$.

Note the difference in the Lebesgue spaces allowed for the coefficients in Theorems 6.1.2 and 6.1.5. In passing we comment that in [43] the spaces are correctly stated on page 276 for the analogue of Theorem 6.1.5, but seem to be too weak for Theorem 6.1.2 in the case $p > n$.

Theorem 6.1.5 applies in particular to the linear elliptic inequality

$$\partial_{x_i}\{a_{ij}(x)\partial_{x_j}u\} + b_i(x)\partial_{x_i}u + c(x)u \geq f(x),$$

provided that the coefficients $[a_{ij}]$, b_i are bounded, the coefficient c is non-positive, and $f \in L^q(\Omega)$ for some $q > n/2$. In fact here

$$B(x, z, \boldsymbol{\xi}) = b_i(x)\boldsymbol{\xi}_i + c(x)z - f(x) \leq b_1|\boldsymbol{\xi}| + |f(x)|,$$

when $z \geq 0$, so the required hypotheses are satisfied with $p = 2$.

For the special case of the p-Laplace inequality

$$\Delta_p u + B(x, u, Du) \geq 0, \quad \text{with } B(x, z, \boldsymbol{\xi}) \leq b_1 |\boldsymbol{\xi}|^{p-1} + b^{p-1}, \quad (6.1.11)$$

the above results can usefully be compared with Theorem 3.7.4, or with Theorem 2.3.2 if $u \in C^2(\Omega)$. Indeed, when $\Omega = \{x \in \mathbb{R}^n : -R < x_1 < R\}$ and $M = 0$, Theorem 3.7.4 gives

$$u(x) \leq (p-1)^{-1/(p-1)} \left[e^{1+b_1 R/(p-1)} - 1 \right] b \, R^{p'} \quad \text{a.e. in } \Omega, \quad (6.1.12)$$

(we have written b^{p-1} for b to facilitate the comparison). On the other hand, when $\Omega = B_R$ we find from Theorem 6.1.4 in the case $a = 0$,

$$u(x) \leq C(p, n, Rb_1) \, b \, R^{p'} \quad \text{a.e. in } B_R. \quad (6.1.13)$$

The estimate $(6.1.12)$ is considerably better than $(6.1.13)$, but of course the class of equations covered by Theorem 6.1.4 contains inequalities not included in Theorem 3.7.4 (and vice versa).

When $b_1 = 0$ the explicit solution

$$u(x) = [n^{-1/(p-1)} b/p'] \cdot (R^{p'} - |x|^{p'}), \qquad x \in B_R,$$

of $(6.1.11)$ shows that the optimal estimate is $u(x) \leq n^{-1/(p-1)} b \, R^{p'}/p'$.

A second point of comparison can be made with the estimate of Alexandrov (Theorem 9.1 of [43]). For simplicity, consider the non-homogeneous Laplace equation $\Delta u + f(x) = 0$ in the ball $\Omega = B_R$, $n \geq 2$, with $u \leq 0$ on ∂B_R. From Theorem 6.1.3 or Theorem 6.1.4, and Theorem 6.1.5, we see that, for $u \in W^{1,2}_{\text{loc}}(B_R)$,

$$u(x) \leq C(n, \varepsilon) R^{\varepsilon/2} \|f\|_{n/2(1-\varepsilon), B_R} \quad \text{a.e. in } B_R.$$

On the other hand, Theorem 9.1 of [43] for this case states that, for $u \in W^{2,n}(B_R)$,

$$u(x) \leq C(n, R) \|f\|_{n, B_R} \quad \text{in } B_R.$$

Clearly the first estimate is better for the case in question. On the other hand, the difference in the range of operators allowed here and in Alexandrov's theory is considerable.

Finally the Lebesgue spaces $(6.1.9)$ are in all probability best possible. For definiteness, consider the p-Laplace operator with $1 < p < n$. One can check that the function $u(x) = [\log(1/|x|)]^\gamma$, $\gamma > 0$, is in $W^{1,p}(B_1)$ and that u is a solution of the equation $\Delta_p u + b^{p-1} = 0$ with $b \in L^{n/p'}(B_1)$, provided only that $n > p \max\{p-1, 1/(p-1)\}$ and $\gamma < [(n-1)p - n]/(n-1)p$, while nevertheless u is unbounded as $x \to 0$.

6.2 Proof of Theorems 6.1.1 and 6.1.2

We begin with two crucial lemmas of independent interest. Their proofs could be treated in more condensed form, but it seems best here to proceed at a more deliberate pace.

Lemma 6.2.1. *Assume $|\Omega| = 1$. Suppose that the functions $\boldsymbol{A}$ and B satisfy (6.1.2) with $1 \leq p < n$, and let $u \in W^{1,p}_{loc}(\Omega)$ be a p-regular solution of the inequality (6.1.1), such that $u \leq 0$ on $\partial\Omega$.*
  Define $w = u^+ + k$, where $k = a + b \geq 0$. Then $w \in L^{\infty}(\Omega)$ and

$$w \leq C[1 + b_1 + (a_2 + b_2)^{1/p}]^{n/p}\|w\|_p \qquad a.e. \text{ in } \Omega;$$

the constant $C = C(p, n)$ can be taken in the specific form $[6(1 + S)]^{n/p}$, with $S = S(p, n)$, the Sobolev constant for the embedding from $W^{1,p}_0(\Omega)$ into $L^{p^*}(\Omega)$, $p^* = np/(n - p)$.

Proof. Clearly $w \geq k$ in Ω, $w = k$ on $\partial\Omega$, and of course $w \in W^{1,p}_{loc}(\Omega)$. *Step 1.* Let ℓ, m be fixed, with $k < \ell < m$ (ultimately we take $\ell \to k$ and $m \to \infty$). Define

$$\psi(t) = \frac{r^p}{q} \begin{cases} 0, & \text{if} \quad t \leq \ell, \\ t^q - \ell^q, & \text{if} \quad \ell < t < m, \\ qm^{q-1}t - (q-1)m^q - \ell^q, & \text{if} \quad t \geq m; \end{cases} \qquad (6.2.1)$$

$$v(t) = \begin{cases} \ell^r, & \text{if} \quad t \leq \ell, \\ t^r, & \text{if} \quad \ell < t < m, \\ rm^{r-1}t - (r-1)m^r, & \text{if} \quad t \geq m, \end{cases} \qquad (6.2.2)$$

where q and r are real parameters, with $q \geq 1$ and r determined by the relation

$$q - 1 = p(r - 1). \qquad (6.2.3)$$

Thus ψ and v are convex, piecewise smooth except for corners at $t = \ell$, and linear when $t \geq m$.
  By Lemma 3.1.2 with $f = w$ and $\ell' = k$, it is clear that $\varphi = \psi(w)$ can serve as a test function for (6.1.1) in Ω. In particular by (3.1.5),

$$\int_{\Omega} \langle \boldsymbol{A}(x, u, Du), D\varphi \rangle \leq \int_{\Omega} [B(x, u, Du)]^+ \varphi. \qquad (6.2.4)$$

When $w \le \ell$ we have $\varphi = 0$, so that the integrals need be evaluated only over the set $\{x \in \Omega : \ell < w(x) < \infty\}$. But in this set necessarily $u(x) > 0$, $u^+ = u$, and in turn

$$u = w - k, \qquad Du = Dw. \tag{6.2.5}$$

Also $D\varphi = \psi'(w)Dw$, so that by (6.1.2) and (6.2.5),

$$\langle \boldsymbol{A}(x, u, Du), D\varphi \rangle \ge \psi'(w)\{|Dw|^p - a_2 w^p - a^p\},$$
$$[B(x, u, Du)]^+ \varphi \le \psi(w)\{b_1|Dw|^{p-1} + b_2 w^{p-1} + b^{p-1}\}. \tag{6.2.6}$$

To evaluate the right sides of (6.2.6) we require some preliminary estimates. First, using the relation (6.2.3) between q and r we have

$$\psi'(t) = [v'(t)]^p,$$

and

$$\psi(t) = \int_0^t [v'(s)]^p ds \le [v'(t)]^{p-1} \int_0^t v'(s) ds = v(t)[v'(t)]^{p-1}.$$

Moreover, using (6.2.2) one finds that $tv'(t) \le rv(t)$. Putting

$$v = v(x) = v \circ w(x),$$

the terms on the right side of (6.2.6) then have the following estimates:

$$\begin{aligned}
\psi'(w)|Dw|^p &= |v'(w)|^p|Dw|^p = |Dv|^p, \\
\psi'(w)w^p &= [wv'(w)]^p \le r^p v^p, \\
\psi(w)w^{p-1} &\le v(w)[wv'(w)]^{p-1} \le r^{p-1}v^p, \\
\psi(w)|Dw|^{p-1} &\le v|v'(w)Dw|^{p-1} = v|Dv|^{p-1},
\end{aligned} \tag{6.2.7}$$

by (6.2.3). This being shown, (6.2.4) now takes the form

$$\int_\Omega |Dv|^p \le \int_\Omega b_1 v|Dv|^{p-1} + r^p \int_\Omega (a_2 + b_2 + c_2)v^p, \tag{6.2.8}$$

where (recalling that $w \ge k$)

$$c_2 = (a/k)^p + (b/k)^{p-1}, \qquad (c_2 = 0 \text{ if } k = 0). \tag{6.2.9}$$

The integrals in (6.2.8) are well defined, since $v \in W^{1,p}(\Omega)$ by (6.2.2). Of course $k > 0$ unless $a = b = 0$, in which case we can take $c_2 = 0$.

This beautiful inequality is the key to the lemma.

Step 2. We need the following two (sub)lemmas.

Lemma 6.2.2. *Let α, $\beta > 0$, and $p \geq 1$. If $z^p \leq \alpha z^{p-1} + \beta$, then also*

$$z \leq \alpha + (p\beta)^{1/p}, \qquad z^p \leq \alpha^p + p\beta.$$

Proof. By Young's inequality

$$\alpha z^{p-1} \leq \alpha^p/p + z^p/p'.$$

Hence $z^p/p \leq \alpha^p/p+\beta$, and the result follows at once (note that $(x+y)^{1/p} \leq x^{1/p} + y^{1/p}$). $\square$

Lemma 6.2.3. *We have $\|v\|_{p^*} < \infty$ and (recall $|\Omega| = 1$)*

$$\|v\|_{p^*} \leq S\|Dv\|_p + \|v\|_p. \tag{6.2.10}$$

Proof. Since $v \equiv \ell^r$ near $\partial\Omega$, then $\|v - \ell^r\|_{p^*} \leq S\|Dv\|_p$ by Sobolev's inequality, Theorem 3.9.1. Also $\|\ell^r\|_{p^*} = \|\ell^r\|_p$ since $|\Omega| = 1$. Therefore

$$\|v\|_{p^*} \leq \|v - \ell^r\|_{p^*} + \|\ell^r\|_{p^*} \leq S\|Dv\|_p + \|\ell^r\|_p$$

and the lemma now follows since $v \geq \ell^r$ in Ω. $\square$

Step 3. By Hölder's inequality

$$\int_\Omega v|Dv|^{p-1} \leq \|v\|_p\|Dv\|_p^{p-1}, \qquad \int_\Omega v^p = \|v\|_p^p.$$

Define

$$z = \|Dv\|_p/\|v\|_{p^*}, \qquad y = \|v\|_p/\|v\|_{p^*},$$

which can be done since $\|v\|_{p^*} \geq \ell^r > 0$. Then from the key formula (6.2.8) there follows, after division by $\|v\|_{p^*}^p$,

$$z^p \leq b_1 y z^{p-1} + cr^p y^p, \tag{6.2.11}$$

where $c = a_2 + b_2 + 2$. To see this we recall that $k = a + b$; hence by the definition (6.2.9) of c_2 there holds $c_2 \leq 2$ and $a_2 + b_2 + c_2 \leq c$.[1]

This being shown, from Lemma 6.2.2 we obtain

$$z \leq [b_1 + (pc)^{1/p}r]y \leq dry, \tag{6.2.12}$$

where $d = b_1 + (pc)^{1/p}$.

---

[1] When $p \geq 2$ or when either a or b is 0, one can take $c = a_2 + b_2 + 1$.

Inequality (6.2.10) can be rewritten in the form

$$1 \le Sz + y. \tag{6.2.13}$$

Consequently, by (6.2.12) we get $1 \le (1 + Sdr)y$. In turn, using the definition of y, there results

$$\|v\|_{p^*} \le (1 + Sdr)\|v\|_p. \tag{6.2.14}$$

The left-hand side of (6.2.14) can be replaced by the smaller norm $\|w\|_{p^*r,\Gamma}^r$, where $\Gamma = \{x \in \Omega : k \le w(x) < m\}$; while on the right the term $\|v\|_p$ can be replaced by the larger one $\|w\|_{pr}^r + (\ell^r - k^r)$ (since $v \le w^r + (\ell^r - k^r)$ and $|\Omega| = 1$). We can now let $\ell \to k$, $m \to \infty$ in this modified version of (6.2.14), yielding (since $\Gamma \nearrow \Omega$)

$$\|w\|_{\kappa pr} \le (Kr)^{1/r}\|w\|_{pr}; \quad \kappa = p^*/p = n/(n-p), \quad K = 1 + Sd \tag{6.2.15}$$

provided however that $w \in L^{pr}(\Omega)$.

We assert that in fact w is in $L^{pr}(\Omega)$ for all $r \ge 1$. This obviously holds for $r = 1$ since the boundary condition for w implies that w is bounded near $\partial\Omega$. Using (6.2.15), an induction argument then proves the assertion. The remarkable inequality (6.2.15) was (in essence) first discovered in the linear homogeneous case by Moser [62].

Step 4. Taking first $r = 1$ in (6.2.15), we get

$$\|w\|_{p\kappa} = \|w\|_{p^*} \le K\|w\|_p.$$

Next, take $r = p^*/p = \kappa$ so that

$$\|w\|_{p\kappa^2} \le (K\kappa)^{1/\kappa}\|w\|_{p\kappa} \le (K\kappa)^{1/\kappa}K\|w\|_p = K^{1+1/\kappa}\kappa^{1/\kappa}\|w\|_p.$$

Continuing in this way, with r successively equal to κ, κ^2, etc., we get

$$\|w\|_{p\kappa^j} \le K^\Sigma \kappa^{\Sigma'}\|w\|_p, \tag{6.2.16}$$

where

$$\Sigma = \Sigma_j = \sum_{i=0}^{j-1} \frac{1}{\kappa^i}, \quad \Sigma' = \Sigma'_j = \sum_{i=1}^{j-1} \frac{i}{\kappa^i}. \tag{6.2.17}$$

The series Σ converges to $\kappa/(\kappa - 1) = n/p$ as $j \to \infty$. Similarly the series Σ' converges to $\kappa/(\kappa - 1)^2 = n(n-p)/p^2$.

Thus letting $j \to \infty$ in (6.2.16) gives

$$\|w\|_\infty \le K^{n/p} \left(\frac{n}{n-p} \right)^{n(n-p)/p^2} \|w\|_p$$

$$= \left[\left(1 + \frac{p}{n-p} \right)^{(n-p)/p} K \right]^{n/p} \|w\|_p \qquad (6.2.18)$$

$$\le (Ke)^{n/p} \|w\|_p.$$

Here, for the record,

$$K = 1 + Sd = 1 + S[b_1 + (pc)^{1/p}] \qquad (6.2.19)$$

$$= 1 + S[b_1 + p^{1/p}(a_2 + b_2 + 2)^{1/p}]. \qquad \square$$

Remark. The proof of Lemma 6.2.1 follows closely the proof of Theorem 1 of [92], with however significant improvements and clarifications of the required calculations.

The special linear case of Lemma 6.2.1 is due to Stampacchia and Maz'ya. The corresponding treatment of this case by Gilbarg and Trudinger [43], Theorems 8.15 and 10.9, is perhaps more concise than necessary.

If one is concerned only with the case $p < n$ and constant values for the coefficients $a_2, \ldots, b$, one can omit the following rather difficult lemma.

Lemma 6.2.4. *Let the hypotheses of Lemma 6.2.1 be satisfied for $p \ge 1$, and assume that the coefficients in (6.1.2) are functions in the respective Lebesgue spaces (6.1.5).*

Let $k = \|a\| + \|b\|$. Then $w = u^+ + k \in L^\infty(\Omega)$ and

$$w \le C[1 + \|b_1\| + \|a_2 + b_2\|^{1/p}]^\nu \|w\|_p, \qquad (6.2.20)$$

where the constant C can be taken in the form $[6(1+\bar{S})]^\nu$, with $\bar{S} = S(s_*, n)$ and

$$s_* = p, \quad \nu = n/\varepsilon p \quad \text{if } p < n; \qquad s_* = \frac{2np}{n\varepsilon + 2p}, \quad \nu = 4/\varepsilon \quad \text{if } p \ge n.$$

We recall that by $\|a\|$, $\|b\|$, $\|b_1\|$ and $\|a_2 + b_2\|$ we mean the norms of a, b, b_1, $a_2 + b_2$ in the respective Lebesgue spaces (6.1.5). Note that the constant C can be quite large.

Proof. We follow the proof of Lemma 6.2.1 but now with the coefficients of (6.1.2) in the respective spaces (6.1.5), and with $k = \|a\| + \|b\|$. For simplicity, the argument will be given in detail only for the case $p > 1$.

Step 1′. With φ and v defined as in the proof of Lemma 6.2.1, and proceeding exactly as before, we obtain again the inequality (6.2.8). Step 2 is next replaced by

Step 2′. Two further lemmas.

Sublemma 1. *Suppose* $\sum_1^j \gamma_i/p_i = 1$. *Then*

$$\int_\Omega \Pi_1^j |f_i|^{\gamma_i} \leq \Pi_1^j \|f_i\|_{p_i}^{\gamma_i}.$$

This is a consequence of Hölder's inequality, though seems not to be explicitly stated in the literature.

Sublemma 2. *Let* $\theta = 1$ *and* $s = p^* = pn/(n-p)$ *if* $1 < p < n$, *while* $\theta = 2$ *and* $s = 2p/\varepsilon$ *if* $p \geq n$. *Then*

$$\int_\Omega b_1 v |Dv|^{p-1} \leq \|b_1\|_{p\alpha} \|v\|_p^{\varepsilon/\theta} \|v\|_s^{1-\varepsilon/\theta} \|Dv\|_p^{p-1},$$

$$\int_\Omega (a_2 + b_2) v^p \leq \|a_2 + b_2\|_\alpha \|v\|_p^{p\varepsilon/\theta} \|v\|_s^{p(1-\varepsilon/\theta)},$$

$$\int_\Omega a^p v^p \leq \|a\|_{p\alpha}^p \|v\|_p^{p\varepsilon/\theta} \|v\|_s^{p(1-\varepsilon/\theta)},$$

$$\int_\Omega b^{p-1} v^p \leq \|b\|_{(p-1)\alpha}^{p-1} \|v\|_p^{p\varepsilon/\theta} \|v\|_s^{p(1-\varepsilon/\theta)}.$$

Proof. When $p < n$ the first line is a direct consequence of Sublemma 1 applied to the *four*-fold product $b_1 v^{\varepsilon/\theta} v^{1-\varepsilon/\theta} |Dv|^{p-1}$. The remaining inequalities for $p < n$ follow in the same way.

When $p \geq n$, one has $\alpha = 1/(1-\varepsilon)$, $\theta = 2$, $s = 2p/\varepsilon$. Then for the first line we use Sublemma 1 with the *five*-fold product $b_1 v^{\varepsilon/\theta} v^{1-\varepsilon/\theta} |Dv|^{p-1} \cdot 1$ and the exponent relation

$$\frac{1}{p\alpha} + \frac{\varepsilon}{p\theta} + \frac{1-\varepsilon/\theta}{s} + \frac{p-1}{p} + \frac{1}{\delta} = 1, \qquad \delta = \frac{4p}{\varepsilon^2}.$$

Since $|\Omega| = 1$ the extra term $\|1\|_\delta$ in the Hölder product in fact does not explicitly appear. The remaining inequalities follow in the same way, with however $\delta = 4/\varepsilon^2$ in these cases. $\square$

From the last three inequalities of the lemma and the fact that $k = \|a\| + \|b\|$ we obtain

$$\int_{\Omega} (a_2 + b_2 + c_2)v^p \le (\|a_2 + b_2\| + 2)\|v\|_p^{p\varepsilon/\theta}\|v\|_s^{p(1-\varepsilon/\theta)}. \qquad (6.2.21)$$

Step 3'. Set

$$z = \|Dv\|_p/\|v\|_s, \qquad y = \|v\|_p/\|v\|_s,$$

where the value of the parameter s is given in Sublemma 2. Then from (6.2.8) we find, using the first inequality of Sublemma 2 together with (6.2.21), that (see (6.2.11))

$$z^p \le \|b_1\| \, y^{\varepsilon/\theta} z^{p-1} + c \, r^p y^{p\varepsilon/\theta},$$

where $c = \|a_2 + b_2\| + 2$.

The rest of the proof is essentially the same as before, with however (6.2.12) being replaced by

$$z \le \{\|b_1\| + (pc)^{1/p} r\} y^{\varepsilon/\theta}.$$

Using Lemma 6.2.3 with p^* replaced by s, and so also $S(p,n)$ replaced by $\bar{S} = S(s_*, n)$, see Theorem 3.9.2, then gives in place of (6.2.13),

$$1 \le \bar{S} d r y^{\varepsilon/\theta} + y,$$

with $d = \|b_1\| + (pc)^{1/p}$. In turn, from Lemma 6.2.2 in the case $z = 1$ and exponent $\theta/\varepsilon \, (\ge 1)$, one gets

$$1 \le [(\bar{S}dr)^{\theta/\varepsilon} + \theta/\varepsilon]y.$$

It now follows that, see (6.2.14),

$$\|v\|_s \le (\bar{K}r)^{\theta/\varepsilon}\|v\|_p,$$

where

$$\bar{K} = (\theta/\varepsilon)^{\varepsilon/\theta} + \bar{S}d \le e^{1/e} + \bar{S}d.$$

Reverting to the variable w we get

$$\|w\|_{\kappa pr} \le (\bar{K}r)^{\theta/\varepsilon r}\|w\|_{pr}, \qquad \kappa = s/p. \qquad (6.2.22)$$

Remark. For the case $p = 1$, where $b_1 = 0$ and b_2 is replaced by b, the calculation is slightly simpler. We can then take $\bar{K} = e^{1/e} + \bar{S}(\|a_2 + b\| + 1)$, or, if $\varepsilon = 1$, even $\bar{K} = 1 + \bar{S}(a_2 + b + 1)$.

Step 4'. The proof is now concluded by iteration, as in the case of Lemma 6.2.1. In fact, when $p < n$ we have $\theta = 1$, $\kappa = n/p^* = n/(n-p)$, so the same calculation used in the derivation of of (6.2.18) gives now

$$\|w\|_\infty \le (Ke)^{n/p\varepsilon}\|w\|_p \tag{6.2.23}$$

with $K = e^{1/e} + \bar{S}[\|b_1\| + p^{1/p}(\|a_2 + b_2\| + 2)^{1/p}]$. When $p \ge n$ the situation is slightly different. In this case $\theta = 2$ and $s = 2p/\varepsilon$, so that $\kappa = s/p = 2/\varepsilon$. The main series Σ and Σ' then converge respectively to $2/(2-\varepsilon)\,(<2)$ and $2\varepsilon/(2-\varepsilon)^2\,(<2)$. Thus we find

$$\|w\|_\infty \le (\bar{K}^\Sigma)^{2/\varepsilon}[(2/\varepsilon)^{\Sigma'}]^{2/\varepsilon}\|w\|_p \le (\bar{K}e^{2/e})^{4/\varepsilon}\|w\|_p \tag{6.2.24}$$

because $(2/\varepsilon)^{\Sigma'/2} \le (2/\varepsilon)^\varepsilon \le e^{2/e}$.

The conclusions (6.2.23) for $1 < p < n$ and (6.2.24) for $p \ge n$ can be combined to give (6.2.20) with the constant $C = [6(1+\bar{S})]^\nu$. This completes the proof.   □

Proof of Theorem 6.1.2. *Step* 1. Consider first the case

$$M = 0, \qquad |\Omega| = 1, \qquad p > 1. \tag{6.2.25}$$

Take $k = \|a\| + \|b\|$ and $w = u^+ + k$. Then by Lemma 6.2.4 we have

$$w \le \text{Const. } \|w\|_p, \tag{6.2.26}$$

where the constant depends on p, n, ε, $|\Omega|$, $\|b_1\|$, $\|a_2 + b_2\|$. Therefore

$$u \le \text{Const.}(\|u^+\|_p + k),$$

which gives (6.1.4) for the case (6.2.25).

Step 2. When $M > 0$, $p > 1$ we first define $\tilde{u} = u - M$, so that $\tilde{u} \le 0$ on $\partial\Omega$. Furthermore $\tilde{u}$ satisfies (6.1.1) with the coefficients a_2, b_2, a, b in (6.1.2) respectively replaced by

$$\begin{aligned}
\tilde{a}_2 &= 2^{p-1}a_2, & \tilde{a} &= a + 2^{1/p'}a_2^{1/p}M, \\
\tilde{b}_2 &= 2^{p-1}b_2, & \tilde{b} &= 2^{p'}(b + b_2^{1/(p-1)}M).
\end{aligned} \tag{6.2.27}$$

Proof. We treat the case of a_2 and a, leaving b_2 and b to the reader. First,

$$a_2u^p = a_2(\tilde{u} + M)^p \le 2^{p-1}a_2(\tilde{u}^p + M^p).$$

Thus $\tilde{a}_2 = 2^{p-1}a_2$, and

$$\tilde{a}^p = a^p + 2^{p-1}a_2M^p \le (a + 2^{1/p'}a_2^{1/p}M)^p,$$

as required by (6.2.27).   □

Now take $\tilde{k} = \|\tilde{a}\| + \|\tilde{b}\|$. Then $\tilde{w} = \tilde{u}^+ + k$ obeys (6.2.26) with the constant depending on p, n, ε, $\|b_1\|$, $\|\tilde{a}_2 + \tilde{b}_2\|$, that is, on p, n, ε, $\|b_1\|$, $\|a_2 + b_2\|$. The conclusion of Theorem 6.1.2 is thus proved subject to the condition $|\Omega| = 1$, $p > 1$.

Step 3. The general case $|\Omega| \neq 1$ is obtained by a change of scale $x = R\bar{x}$, with $R = |\Omega|^{1/n}$ so that $|\bar{\Omega}| = 1$. In the new scale $\bar{A} = R^{-1}A$ and $\bar{B} = B$, or equivalently $\bar{A} = R^{p-1}A$, $\bar{B} = R^p B$. In turn a, b, b_1, a_2, b_2 in (6.1.2) are replaced in the new scale by

$$|\Omega|^{1/n}a, \quad |\Omega|^{p'/n}b, \quad |\Omega|^{1/n}b_1, \quad |\Omega|^{p/n}a_2, \quad |\Omega|^{p/n}b_2,$$

while the norms $\|a\|$, $\|b\|$, $\|b_1\|$ and $\|a_2 + b_2\|$ are correspondingly replaced by

$$|\Omega|^{\gamma}\|a\|, \quad |\Omega|^{p'\gamma}\|b\|, \quad |\Omega|^{\gamma}\|b_1\|, \quad |\Omega|^{p\gamma}\|a_2 + b_2\|,$$

$$\gamma = \frac{1}{n} - \frac{1}{p\alpha} = (1 - \varepsilon)\left[\frac{p - n}{pn}\right]^+ + \frac{\varepsilon}{n}, \tag{6.2.28}$$

with α defined in (6.1.5).

Step 4. Finally, if $p = 1$ then b_1, b_2 are dropped from (6.2.28), while b is replaced by $|\Omega|^{1/n}b$ and $\|b\|$ by $|\Omega|^{\varepsilon/n}\|b\|$. Moreover, since b replaces b_2 and b^{p-1} is discarded, we take $k = \|a\|$ with the constant C in (6.1.4) depending on n, ε, $|\Omega|$, $\|a_2 + b\|$; see the note at the end of Step 3'. □

Theorem 6.1.1 is obtained from the special case $\varepsilon = 1$.

6.3 Proof of Theorem 6.1.3 and the first part of Theorem 6.1.5

Lemma 6.3.1. *Let the hypotheses of Lemma 6.2.1 hold, with $|\Omega| = 1$ and with the additions that $u^+ \in L^{\infty}(\Omega)$, $1 < p \leq n$, and $b_1 = b_2 = 0$. Assume the coefficients a, b, a_2 are in the respective Lebesgue spaces (6.1.5). Suppose also $k = \|a\| + \|b\| > 0$ and $\|w\|_p \geq 2k$, where $w = u^+ + k$. Then*

$$\log \frac{W}{k} \leq 2\{1 + Q(\|a_2\| + p')^{1/p}\}\frac{W}{\|w\|_p},$$

where $W = \|w\|_{\infty}$ and $Q = w_n^{-1/n}$ is Poincaré's constant (Theorem 3.9.4).

Proof. It is enough to treat only the non-trivial case $k < W$. Let ℓ be an arbitrary constant, with $\ell \in (k, W)$, and define

$$\psi(t) = \begin{cases} 0, & \text{if} \quad t \leq \ell, \\ \ell^{1-p} - t^{1-p}, & \text{if } \ell < t \leq W. \end{cases}$$

We choose $\varphi = \psi(w)$ as test function for (6.1.1). Putting $\Gamma = \{x \in \Omega : \ell < w(x) \leq W\}$, then $\varphi = 0$, $D\varphi = \mathbf{0}$ in $\Omega \setminus \Gamma$ and $D\varphi = (p-1)w^{-p}Dw$ in Γ. Therefore from (3.1.5) and (6.1.2) we get

$$(p-1) \int_{\Gamma} w^{-p} [|Dw|^p - a_2 w^p - a^p] \leq \int_{\Gamma} (b/\ell)^{p-1}. \qquad (6.3.1)$$

Also

$$\int_{\Gamma} \left[(p-1) \left(\frac{a}{k} \right)^p + \left(\frac{b}{k} \right)^{p-1} \right] \leq (p-1) \left(\frac{\|a\|}{k} \right)^p + \left(\frac{\|b\|}{k} \right)^{p-1} \leq p.$$

Therefore, since $w > \ell > k$ in Γ, the inequality (6.3.1) yields

$$(p-1) \int_{\Gamma} |D \log w|^p \leq (p-1)\|a_2\|_1 + p. \qquad (6.3.2)$$

By Poincaré's inequality (note that $\log(w/\ell) \in W^{1,p}(\Omega)$, $\log(w/\ell) = 0$ in $\Omega \setminus \Gamma$ and $|\Omega| = 1$)

$$\| \log(w/\ell)\|_p \leq Q\|D \log(w/\ell)\|_p \leq Q(\|a_2\| + p')^{1/p}. \qquad (6.3.3)$$

But $1 < w/\ell \leq W/\ell$ in Γ, whence

$$w \leq \frac{W}{1 + \log(W/\ell)} \left(1 + \log \frac{w}{\ell} \right) \qquad \text{in } \Gamma.$$

By integration

$$\|w\|_{p,\Gamma} \leq \frac{W}{1 + \log(W/\ell)} \left(1 + \left\| \log \frac{w}{\ell} \right\|_{p,\Gamma} \right)$$

$$\leq \{1 + Q(\|a_2\| + p')^{1/p}\} \frac{W}{1 + \log(W/\ell)}.$$

Next observe that $\|w\|_p = \|w\|_{p,\Omega^+} + k|\Omega|$, where $\Omega^+ = \{x \in \Omega : w(x) > k\}$. Thus, since $|\Omega| = 1$ and $\|w\|_p \geq 2k$ by assumption, we get

$\|w\|_p \leq \|w\|_{p,\Omega^+} + \frac{1}{2}\|w\|_p$, that is $\|w\|_p \leq 2\|w\|_{p,\Omega^+}$. Letting $\ell \to k$ so that $\|w\|_{p,\Gamma} \to \|w\|_{p,\Omega^+}$, it follows that

$$\|w\|_p \leq 2\|w\|_{p,\Omega^+} \leq 2\{1 + Q(\|a_2\| + p')^{1/p}\}\frac{W}{1 + \log(W/k)}.$$

Rearranging proves the lemma. □

Lemma 6.3.2. *Let the hypotheses of Lemma 6.3.1 be satisfied, with the exception that $p > n$ and we no longer assume a priori that $u^+ \in L^\infty(\Omega)$. If $k > 0$ then $w \in L^\infty(\Omega)$ and*

$$\log\frac{W}{k} \leq Q_\infty(\|a_2\| + p')^{1/p},$$

where the constant Q_∞, Morrey's constant, depends only on p and n.

Proof. The inequality (6.3.2) holds equally when $p > n$. The lemma is then an immediate consequence of Theorem 3.9.3. □

Proof of Theorem 6.1.5 when $b_1 = b_2 = 0$. First suppose $M = 0$, $|\Omega| = 1$ and $k = \|a\| + \|b\| > 0$.

Case 1. $\|w\|_p < 2k$. From Lemma 6.2.4 in the case $1 < p < n$ we get

$$w \leq C(1 + \|a_2\|)^{n/\varepsilon p^2}\|w\|_p \leq 2C(1 + \|a_2\|)^{n/\varepsilon p^2}k, \qquad (6.3.4)$$

where the constant C depends only on p, n and ε. Since $w = u^+ + k$ it follows that (6.1.10) holds for this case, that is $u \leq C(\|a\| + \|b\|)$.

Case 2. $\|w\|_p \geq 2k$. By Lemma 6.3.1,

$$\log\frac{w}{k} \leq C(1 + \|a_2\|)^{1/p + n/\varepsilon p^2}$$

(new constant C), and so

$$u \leq k\exp\{C(1 + \|a_2\|)^{(n+\varepsilon p)/\varepsilon p^2}\}.$$

When $p > n$ we apply Lemma 6.3.2 and the conclusion follows as before, using (6.2.24). The proof for the case $p = n$ is essentially the same.

When $M > 0$ the argument is the same as for the proof of Theorem 6.1.2. If $k = 0$ then we replace k by ℓ and let ℓ go to zero. Finally the case $|\Omega| \neq 1$ is treated by a change of scale as in the proof of Theorem 6.1.2. □

Theorem 6.1.3 is obtained from the special case $\varepsilon = 1$.

Remark. That the coefficients are in different Lebesgue spaces in (6.1.9) when $1 < p \le n$ and $p > n$ is due to the use of Lemma 6.2.4 in obtaining (6.3.4) when $p \le n$, a use which is not required when $p > n$.

6.4 Proof of Theorem 6.1.4 and the second part of Theorem 6.1.5

Lemma 6.4.1. *Let the hypotheses of Lemma 6.3.1 be satisfied, with the exception that $a_2 = b_2 = 0$. Suppose $k = \|a\| + \|b\| > 0$ and define (without confusion)*

$$v = \log \frac{W}{W - w + k}, \qquad W = \|w\|_\infty, \qquad w = u^+ + k.$$

Then $v \in W^{1,p}(\Omega) \cap L^\infty(\Omega)$ and

$$\|v\|_p \le \frac{Q}{p-1}(\|b_1\| + 2p). \tag{6.4.1}$$

Moreover v satisfies an inequality of the form

$$\operatorname{div} \bar{A}(x, v, Dv) + \bar{B}(x, v, Dv) \ge 0 \qquad in\ \Omega, \tag{6.4.2}$$

with condition (6.1.2) now valid with A, B, a_2, b_2, a and b replaced respectively by

$$\bar{A}, \ \bar{B}, \ 0, \ 0, \ \bar{a}, \ \bar{b}, \quad where \quad \bar{a} = a/k, \quad \bar{b} = (p-1)(a/k)^p + (b/k)^{p-1}. \tag{6.4.3}$$

Proof. Step 1. Let $\ell \in (k, W)$, and define

$$\psi(t) = \begin{cases} 0, & \text{if } k \le t \le \ell, \\ (W - t + \ell)^{1-p} - W^{1-p}, & \text{if } \ell < t \le W. \end{cases}$$

Clearly $\varphi = \psi(w)$, where $w = u^+ + k$, can be used as a test function for (6.1.1). Moreover,

$$D\varphi = \begin{cases} 0, & \text{in } \Omega \setminus \Gamma, \\ (p-1)(W - w + \ell)^{-p} Dw, & \text{in } \Gamma, \end{cases}$$

where $\Gamma = \{x \in \Omega \;:\; \ell < w \le W\}$. By the usual calculations, using (6.1.2) with $a_2 = b_2 = 0$, we thus obtain

$$
(p-1)\int_\Gamma (W - w + \ell)^{-p}(|Dw|^p - a^p)
$$

$$
\le \int_\Gamma (W - w + \ell)^{1-p}(b_1|Dw|^{p-1} + b^{p-1}). \tag{6.4.4}
$$

Recalling that $W - w + \ell \ge \ell > k$, this leads to

$$
(p-1)\|D\log(W - w + \ell)\|_p^p
$$

$$
\le \int_\Omega [b_1|D\log(W - w + \ell)|^{p-1} + \bar{b}^{p-1}]. \tag{6.4.5}
$$

For convenience, let $\bar{v}$ be the function v with k replaced by ℓ. Then (6.4.5) takes the form

$$
(p-1)\|D\bar{v}\|_p^p \le \int_\Omega (b_1|D\bar{v}|^{p-1} + \bar{b}^{p-1}). \tag{6.4.6}
$$

Here, see (6.1.9),

$$
\begin{aligned}
\|\bar{b}^{p-1}\|_1 &\le \|\bar{b}^{p-1}\|_\beta \\
&\le (p-1)\|(a/k)^p\|_\beta + \|(b/k)^{p-1}\|_\beta \\
&\le (p-1)(\|a\|/k)^p + (\|b\|/k)^{p-1} \\
&\le p
\end{aligned} \tag{6.4.7}
$$

since $k = \|a\| + \|b\|$ and $\|a\| = \|a\|_{p\beta}$, $\|b\| = \|b\|_{(p-1)\beta}$. Also

$$
\int_\Omega b_1|D\bar{v}|^{p-1} \le \|b_1\|_p\|D\bar{v}\|_p^p. \tag{6.4.8}
$$

From (6.4.6)–(6.4.8) we obtain, with the help of Lemma 6.2.2 and Poincaré's inequality,

$$
\|\bar{v}\|_p \le Q\|D\bar{v}\|_p \le Q\left\{ \frac{\|b_1\|_p}{p-1} + \left(\frac{p^2}{p-1}\right)^{1/p}\right\} \le \frac{Q}{p-1}(\|b_1\| + 2p)
$$

(the constant 2 is an upper bound for the function $I(p) = [(p-1)/p]\cdot[p^2/(p-1)]^{1/p}$; it is easily obtained by writing $I(p) = [(p-1)/p]^{(1-1/p)}\cdot p^{1/p} \le e^{1/e} \approx 1.445$. Letting $\ell \to k$ now proves (6.4.1).

Step 2. We use an ingenious idea of Gilbarg and Trudinger ([43], page 274). Let η be a non-negative test function for (6.1.1) in Ω. Define

$$\psi(t) = (W - t + k)^{1-p}, \qquad k \le t \le W,$$

and take $\varphi = \eta\psi(w)$. Then since ψ, ψ' are bounded in $[k, W]$ it follows that $\varphi \in W^{1,p}(\Omega) \cap L^\infty(\Omega)$, with $\varphi = 0$ near $\partial\Omega$. Hence by a simple extension of Lemma 3.1.2 one can take φ as a (non-negative) test function for (6.1.1). Write $\mu = W - w + k$ and observe that $D\psi(w) = (p-1)\mu^{-p}Dw$. Then since $\varphi = 0$, $D\varphi = \mathbf{0}$ a.e. in the set where $w = k$, we have for all $x \in \Omega$,

$$
\begin{aligned}
\langle \boldsymbol{A}(x, &u, Du), D\varphi \rangle - [B(x, u, Du)]^+\varphi \\
&= \langle \boldsymbol{A}(x, w - k, Dw), D\varphi \rangle - [B(x, w - k, Dw)]^+\varphi \\
&= \mu^{1-p}\langle \boldsymbol{A}(x, w - k, Dw), D\eta \rangle + (p-1)\mu^{-p}\langle \boldsymbol{A}(x, w - k, Dw), Dw \rangle\eta \\
&\quad - \mu^{1-p}[B(x, w - k, Dw)]^+\eta \\
&\ge \mu^{1-p}\langle \boldsymbol{A}(x, w - k, Dw), D\eta \rangle - [b_1(|Dw|/\mu)^{p-1} + \bar{b}^{p-1}]\eta,
\end{aligned}
$$

where we have used (6.1.2) at the last step, along with the inequality $\mu \ge k$ and the definition of $\bar{b}$. Integrating over Ω and using (3.1.5) yields

$$\int_\Omega \{\langle \bar{\boldsymbol{A}}(x, v, Dv), D\eta \rangle - \bar{B}(x, v, Dv)\eta\} \le 0,$$

where

$$
\begin{aligned}
\bar{\boldsymbol{A}}(x, v, Dv) &= \mu^{1-p}\boldsymbol{A}(x, w - k, Dw), \\
\bar{B}(x, v, Dv) &= b_1(|Dw|/\mu)^{p-1} + \bar{b}^{p-1} = b_1|Dv|^{p-1} + \bar{b}^{p-1},
\end{aligned} \qquad (6.4.9)
$$

and we have used the relations $w - k = (1 - e^{-v})W$ and $Dw = We^{-v}Dv = \mu Dv$.

We claim that (6.1.2) holds for $\bar{\boldsymbol{A}}$, $\bar{B}$ with a_2, b_2, a, b respectively replaced by $0, 0, \bar{a}, \bar{b}$. Indeed, again since $a_2 = 0$,

$$\langle \bar{\boldsymbol{A}}(x, v, Dv), Dv \rangle = \mu^{-p}\langle \boldsymbol{A}(x, w - k, Dw), Dw \rangle \ge |Dv|^p - (a/k)^p,$$

proving the claim and the lemma. □

Lemma 6.4.2. *Let the hypotheses of Lemma 6.3.2 be satisfied, with the exception that $a_2 = b_2 = 0$. Then $w \in L^\infty(\Omega)$ and*

$$v = \log \frac{W}{W - w + k} \le \frac{Q_\infty}{p-1}(\|b_1\| + 2p), \qquad W = \|w\|_\infty.$$

Proof. Inequality (6.4.1) holds equally when $p > n$. The lemma is then an immediate consequence of Theorem 3.9.3.   □

Proof of Theorem 6.1.5 when $a_2 = b_2 = 0$. First suppose $1 < p \le n$, $M = 0$, $|\Omega| = 1$ and $k = \|a\| + \|b\| > 0$.

Put $w = u^+ + k$ as in the proof of Theorem 6.1.2. Then Lemma 6.4.1 applies, that is $v \in W^{1,p}(\Omega) \cap L^\infty(\Omega)$ satisfies (6.4.2) with $\bar{a}_2 = \bar{b}_2 = 0$ and $\bar{a}, \bar{b}$ given in (6.4.3). Therefore by Theorem 6.1.2 (!) we get

$$v \le C_1(\|v\|_p + \bar{k}) \tag{6.4.10}$$

with $\bar{k} = \|\bar{a}\| + \|\bar{b}\|$ and $C_1 = C(p, n, \varepsilon)(1 + \|b_1\|)^{n/p}$. On the other hand, from (6.4.7) one has $\bar{k} = \|a\|/k + \|\bar{b}^{p-1}\|_\beta^{1/(p-1)} \le 1 + p^{1/(p-1)} \le 1 + e$.

Then by (6.4.10), together with (6.4.1) and the definition of v, one obtains (a.e. in Ω)

$$\log \frac{W}{W - w + k} \le \frac{C_1 Q}{p - 1}(\|b_1\| + 2p) + (1 + e)C_1$$
$$\le C(p, n, \varepsilon)(1 + \|b_1\|)^{1+n/p} \equiv D.$$

Solving for w we have

$$w \le W(1 - e^{-D}) + k \qquad \text{a.e. in } \Omega,$$

and in turn $W \le k e^D$ since essup $w = W$. The required conclusion (6.1.8) now follows for the case in hand, that is $u \le C(\|a\| + \|b\|)$.

When $p > n$, $M = 0$, $|\Omega| = 1$, $k > 0$ we obtain directly from Lemma 6.4.2 that

$$\log \frac{W}{W - w + k} \le \frac{Q_\infty}{p - 1}(\|b_1\| + 2p)$$

and the conclusion again follows.

To remove the conditions $M = 0$, $|\Omega| = 1$, $k > 0$ we proceed exactly as in the earlier proof of Theorem 6.1.5 for the case $b_1 = b_2 = 0$. (Since now, however, $a_2 = b_2 = 0$ there is no need to invoke (6.2.27).)   □

Remark. For constant coefficients the previous arguments could be simplified by taking $\varepsilon = 1$ throughout. On the other hand the results for $\varepsilon \in (0, 1)$ seem needed in order to justify the Lebesgue spaces (6.1.9) asserted (without proof) in [43].

6.5 The case $p = 1$ and the mean curvature equation

For the case $p = 1$ the structure conditions (6.1.3) become, for all $(x, z, \xi) \in \Omega \times \mathbb{R}^+ \times \mathbb{R}^n$,

$$|A(x, z, \xi)| \leq \text{Constant}, \qquad \langle A(x, z, \xi), \xi \rangle \geq |\xi| - cz - a, \tag{6.5.1}$$
$$B(x, z, \xi) \leq b$$

(the previous coefficient a_2 is here called c for simplicity).

For this behavior, it is apparent that Theorem 6.1.3 cannot hold without modification. To obtain a corresponding result, we first give a counterpart of Lemma 6.2.4.

Lemma 6.5.1. *Let $u \in W^{1,1}_{\text{loc}}(\Omega)$ be a distribution solution of inequality (6.1.1) with $u \leq 0$ on $\partial\Omega$. Suppose the coefficients a, b, c in (6.5.1) are in the Lebesgue space $L^q(\Omega)$ for some $q > n$.[2]*

Assume that $n \geq 1$, $|\Omega| = 1$ and define $w = u^+ + k$ with $k = \alpha S\|a\|_q$, $\alpha > 0$. Then $w \in L^\infty(\Omega)$ and

$$w \leq \left[\left(\frac{n}{n-1} \right)^{n-1} K \right]^{nq/(q-n)} \|w\|_1, \tag{6.5.2}$$
$$K = e^{1/e} + 1/\alpha + S\|b + c\|_q;$$

here

$$S = S(1, n) = n^{-1}\omega_n^{-1/n}, \qquad \omega_n = \frac{\pi^{n/2}}{\Gamma(1 + n/2)},$$

is the Sobolev constant for $W^{1,1}_0(\mathbb{R}^n)$. In the case of constant coefficients we can take $K = 1 + 1/\alpha + S(b + c)$.

Remark. For the results of this section, the condition that u be 1-regular is redundant by (6.5.1).

Proof of Lemma 6.5.1. We follow the proof of Lemma 6.2.4, but using more precise constants. Indeed, since $p = 1$ we have $\theta = 1$, $s = 1^* = n/(n-1)$. Then writing $q = n/(1-\varepsilon)$, that is $\varepsilon = (q-n)/q$, one gets, in place of (6.2.22),

$$\|w\|_{rn/(n-1)} \leq (Kr)^{q/(q-n)r}\|w\|_r;$$

---

[2] The value q here should not be confused with the parameter in (6.2.1).

the best value for K is given by the expression for $\bar{K}$ in the remark after (6.2.22), with $\bar{S} = S(1, n)$ and the quantity $\|a_2 + b\| + 1$ replaced by

$$\|c + b\| + \frac{\|a\|}{k} = \|b + c\| + \frac{1}{S\alpha}.$$

Finally, as in the iteration step (6.2.23) we obtain (6.5.2); here rather than the constant e in (6.2.23), which holds for *all* $n \geq 1$, we use the precise value $[n/(n-1)]^{n-1}$, see (6.2.18). □

The case $n = 1$ is allowed in this result since $S(1, 1)$ is finite. Also in (6.5.2) the expression inside the brackets reduces simply to K when $n = 1$.

Lemma 6.5.2. *Let the hypotheses of Lemma 6.5.1 hold with the exception that*

$$\|b + c\|_n \leq (1 - \delta)/S, \tag{6.5.3}$$

where $0 < \delta < 1$. Then

$$\|w\|_1 \leq (1 + \alpha)S\|a\|/\delta. \tag{6.5.4}$$

Proof. As in Step 1 of the proof of Lemma 6.2.1, but using only the case $p = 1$, $q = r = 1$, we obtain corresponding to (6.2.8),

$$\|Dv\|_1 \leq \int_\Omega [(b + c)v + a]$$

(application of the inequality $a \leq c_2 v$ is not needed!). In turn, by Hölder's inequality,

$$\|Dv\|_1 \leq \|b + c\|_n \|v\|_{n/(n-1)} + \|a\|_1 \leq \frac{1 - \delta}{S}\|v\|_{n/(n-1)} + \|a\|_1 \tag{6.5.5}$$

by (6.5.3).

Next by Sobolev's inequality,

$$\|v\|_{n/(n-1)} \leq \|v - \ell\|_{n/(n-1)} + \|\ell\|_{n/(n-1)} \leq S\|Dv\|_1 + \ell.$$

Using (6.5.5) this gives

$$\|v\|_{n/(n-1)} \leq (1 - \delta)\|v\|_{n/(n-1)} + S\|a\|_q + \ell.$$

Here one can take $\ell \to k$, $m \to \infty$. Then $v \to w$ (recall $v = w$ when $w > \ell$), from which follows

$$\|w\|_1 \leq \|w\|_{n/(n-1)} \leq (1 + \alpha)S\|a\|_q/\delta. \tag{6.5.6}$$

□

Theorem 6.5.3. *Let* $u \in W_{\mathrm{loc}}^{1,1}(\Omega)$ *be a solution of inequality* (6.1.1) *in* Ω, *with* $\boldsymbol{A}$, B *satisfying* (6.5.1) *and with* a, b, c *in the Lebesgue space* $L^q(\Omega)$ *for some* $q > n$.

Assume that

$$\left\| \frac{b+c}{n} \right\|_q \cdot |\Omega|^{1/n-1/q} < \omega_n^{1/n}, \tag{6.5.7}$$

and suppose $u \leq 0$ *on* $\partial\Omega$. *Then* $u^+ \in L^\infty(\Omega)$ *and*

$$u \leq Ca/\delta, \qquad C = 2S \left[4 \left(\frac{n}{n-1} \right)^{n-1} \right]^n |\Omega|^{1/n} \tag{6.5.8}$$

for constant coefficients, and otherwise

$$u \leq C\|a\|_q/\delta, \qquad C = 3S \left[3 \left(\frac{n}{n-1} \right)^{n-1} \right]^{nq/(q-n)} |\Omega|^{1/n-1/q} \tag{6.5.9}$$

Here $\delta \in (0,1)$ *is a constant such that* (6.5.7) *holds with the right-hand side replaced by* $1 - \delta$.

Proof. First take $|\Omega| = 1$. Define $w = u^+ + k$, $k = \alpha S\|a\|$. Then Lemmas 6.5.1 and 6.5.2 apply. From (6.5.7) and Hölder's inequality we get

$$\|b+c\|_n < \|b+c\|_q < 1/S.$$

Hence

$$K = e^{1/e} + 1/\alpha + S\|b+c\|_q \leq 1 + e^{1/e} + 1/\alpha \tag{6.5.10}$$

by (6.5.3). We are free to choose α as we wish. For constant coefficients take, say, $\alpha = 1$, in which case $1 + \alpha = 2$, $K = 4$, while otherwise take $\alpha = 2$, giving $1 + \alpha = 3$, $K \leq 3$.

The conclusions (6.5.8) and (6.5.9) for the case $|\Omega| = 1$ now follow at once from (6.5.2) and (6.5.6). The general result is then obtained by scaling, cf. relations (6.2.28). $\qquad\square$

Remarks. The key condition (6.5.7) can be replaced by the more elegant inequality

$$\frac{1}{\omega_n} \int_\Omega \left(\frac{b+c}{n} \right)^n < 1,$$

but at the cost that the calculation (6.5.10) no longer applies. Thus (6.5.8) takes the less precise form

$$u \le C \|a\|, \qquad C = C(n, q, |\Omega|, \|b + c\|_n, \|b + c\|_q), \qquad (6.5.11)$$

with the constant C becoming infinite when $q \to n$ or $\|b + c\|_q \to \infty$.

The estimates (6.5.8) and (6.5.11) can be compared with the case $p = 1$ of Theorem 10.10 of [43]. A boundary condition $u \le M$ on $\partial\Omega$ can be handled as earlier, by the change of variable $\tilde{u} = u - M$ and replacing a by $a + M$.

Example. Consider the mean curvature equation

$$\operatorname{div}\left(\frac{Du}{\sqrt{1 + |Du|^2}}\right) = n\mathcal{H}(x). \qquad (6.5.12)$$

Putting $\boldsymbol{A}(\boldsymbol{\xi}) = \boldsymbol{\xi}/\sqrt{1 + |\boldsymbol{\xi}|^2}$, $B(x, u, \boldsymbol{\xi}) = -n\mathcal{H}(x)$, we observe that

$$\langle \boldsymbol{A}(\boldsymbol{\xi}), \boldsymbol{\xi}\rangle \ge |\boldsymbol{\xi}| - a, \qquad a = \sqrt{\frac{5\sqrt{5} - 11}{2}} \sim 0.3002831$$

[this is an easy exercise in differential calculus[3]]. Thus (6.5.12) satisfies (6.5.1), with a given above, $c = 0$, $b = n|\mathcal{H}^-|$ and $|\boldsymbol{A}| < 1$.

From Theorem 6.5.3 we get

$$u \le \left[3\left(\frac{n}{n-1}\right)^{n-1}\right]^{nq/(q-n)} S|\Omega|^{1/n}/\delta \qquad (6.5.13)$$

provided $\|\mathcal{H}^-\|_q|\Omega|^{1/n-1/q} \le (1-\delta)\omega_n^{1/n}$. It is convenient to rewrite this in terms of the *effective radius* of Ω, defined by $|\Omega| = \omega_n\mathcal{R}^n$. Thus it becomes

$$u \le \frac{1}{n}\left[3\left(\frac{n}{n-1}\right)^{n-1}\right]^{nq/(q-n)} \mathcal{R}/\delta,$$

or, when $\mathcal{H}$ is constant,

$$u \le \frac{2a}{n}\left[4\left(\frac{n}{n-1}\right)^{n-1}\right]^{n} \mathcal{R}/\delta.$$

---

[3]The best value for a is $\max_{0 \le t < \infty}(t - t^2/\sqrt{1 + t^2})$. By elementary calculus the maximum occurs at $t_0 = \sqrt{(\sqrt{5} - 1)/2} \sim 0.78614$, and in turn $(t_0 - t_0^2/\sqrt{1 + t_0^2})$ takes the value a given above

For the canonical case when $\mathcal{H}$ is a negative constant and $n = 2$ the second estimate yields $u \leq 10.8\,\mathcal{R}/\delta$ when $|\mathcal{H}|\,\mathcal{R} \leq 1 - \delta$. Of course, this is not too accurate for solutions of class $C^2(\Omega)$ in balls – for which the optimal estimate can be obtained from the elementary maximum principle, using a spherical cap as comparison function. On the other hand, (6.5.13) applies for general domains with finite measure, for solutions in $W_0^{1,1}(\Omega)$ and for $\mathcal{H}^-$ in $L^q(\Omega)$, and provides an explicit upper bound in these cases.

Remark. If $u \leq 0$ on $\partial\Omega$ and $\mathcal{H} = 0$ one expects the conclusion $u \leq 0$ in Ω. This however cannot be obtained by the present approach since the constant $a \sim 0.3002831$ acts as an inhomogeneous term in the structure (6.5.1).

Notes

The semi-maximum principle given in Theorems 6.1.1 and 6.1.2 is due to Serrin (Theorem 3 of [92]), based on earlier work for homogeneous linear equations by Stampacchia [106], Maz'ya [57] and, particularly, Moser [62]. Lemmas 6.3.1 and 6.4.1 for the case of constant coefficients were given by Gilbarg and Trudinger (cases (i) and (ii) on pages 273–274 of [43]). Theorems 6.1.3–6.1.5 are combined work of Gilbarg, Serrin , and Trudinger. Theorems 6.1.3 and 6.1.4 (constant coefficients) were first stated in Theorem 9.7 of [43]. The proofs given here include improved formulations of earlier arguments, e.g., (Lemmas 6.2.4; 6.3.1 and 6.3.2; 6.4.1 and 6.4.2).

The results of Section 6.5 are for the most part new, extending Theorem 10.10 of Gilbarg and Trudinger.

Problems

6.1 Check that the function $u(x) = [\log(1/|x|)]^\gamma$, $\gamma > 0$, is in $W^{1,p}(B_1)$ and that u is a solution of the equation $\Delta_p u + b^{p-1} = 0$ with $b \in L^{n/p'}(B_1)$, provided only that $n > p \max\{p - 1, 1/(p - 1)\}$ and $\gamma < [(n - 1)p - n]/(n - 1)p$.

6.2 Supply the details for the proof of (6.2.28).

6.3 Supply the details for Step 4 in the proof of Theorem 6.1.2.

6.4 Supply the details for example (6.5.12).

6.5 Let u be a p-regular solution of (6.1.1), (6.1.2), with $u \leq 0$ on $\partial\Omega$. Suppose also that Ω is so small that the condition (3.3.1) is satisfied. Then with a, b in the spaces indicated in (6.1.5), show that

$$u \leq C(p, n, a_2, b_1, b_2, |\Omega|)(\|a\| + \|b\|) \qquad \text{a.e. in } \Omega.$$

(Cf. [43], Theorem 10.10.)

[*Hint.* Use Lemma 3.6.2, together with the argument of Theorem 3.3.1 to estimate $\|w\|_p$.]

Chapter 7

The Harnack Inequality

7.1 Local boundedness and the weak Harnack inequality

The ideas of Section 6.2 have far-reaching extensions to questions of local boundedness of solutions of the inequality (6.1.1) and to both weak and strong Harnack-type theorems.[1] These results have already seen application in Section 2.5, but are crucial as well for regularity and existence theory for quasilinear elliptic equations.

The purpose of this section is to present full proofs of these foundational results. For this it is necessary to add to the main conditions (6.1.2) an additional structural inequality, namely

$$|\boldsymbol{A}(x, z, \boldsymbol{\xi})| \le a_1 |\boldsymbol{\xi}|^{p-1} + \bar{a}_2 z^{p-1} + \bar{a}^{p-1}, \tag{7.1.1}$$

and, for the weak Harnack inequality, also

$$B(x, z, \boldsymbol{\xi}) \ge -b_1 |\boldsymbol{\xi}|^{p-1} - b_2 z^{p-1} - b^{p-1}. \tag{7.1.2}$$

The coefficients a, b, b_1, a_2, b_2 are assumed to be in the respective Lebesgue spaces (6.1.5), where $\alpha = n/p(1 - \varepsilon)$, and the coefficients a_1, $\bar{a}$, $\bar{a}_2$ are

---

[1]The idea for Harnack inequalities arises from the famous relation

$$\frac{R - |x|}{(R + |x|)^{n-1}} R^{n-1} u(0) \le u(x) \le \frac{R + |x|}{(R - |x|)^{n-1}} R^{n-1} u(0), \qquad |x| < R,$$

for the values $u(x)$ of a non-negative harmonic function u in a ball of radius R about the origin, first obtained by Axel Harnack in 1887 [45], page 62.

respectively a positive constant (≥ 1) and functions in the Lebesgue spaces $L^n(\Omega)$, $L^{n/(p-1)}(\Omega)$.

Theorem 7.1.1 (Local Boundedness). *Let $u \in W^{1,p}_{\mathrm{loc}}(\Omega)$, $1 \leq p < n$, be a solution of the inequality (6.1.1), with the functions $\boldsymbol{A}$ and B satisfying (6.1.2) and (7.1.1).[2] Then for any open ball B_{2R} in Ω and any $s > p - 1$ we have*

$$\sup_{B_R} u \leq C \left[R^{-n/s} \|u^+\|_{s,B_{2R}} + k(R) \right],$$

where

$$k(R) = R^\varepsilon \|a\| + R^{p'\varepsilon} \|b\| + \|\bar{a}\| \tag{7.1.3}$$

and the constant C depends only on p, n, s, ε; a_1, $\|\bar{a}_2\|$, $R^\varepsilon \|b_1\|$ and $R^{p\varepsilon} \|a_2 + b_2\|$.

Theorem 7.1.2 (Weak Harnack Inequality). *Let $u \in W^{1,p}_{\mathrm{loc}}(\Omega)$, $1 < p < n$, be a non-negative solution of the (reverse) inequality*

$$\mathrm{div}\,\boldsymbol{A}(x, u, Du) + B(x, u, Du) \leq 0. \tag{7.1.4}$$

Suppose that $\boldsymbol{A}$ and B satisfy the first inequality of (6.1.2) together with (7.1.1), (7.1.2). Then for any ball B_{4R} in Ω and any $s \in (0, (p-1)n/(n-p))$ we have

$$R^{-n/s} \|u\|_{s,B_{2R}} \leq C \left[\inf_{B_{2R}} u + k(R) \right],$$

where C and $k(R)$ are as in Theorem 7.1.1.

By sup and inf we mean here essup and essinf. Note that Theorem 7.1.1 shows in particular that u is bounded above on any compact subset of Ω. Also Theorems 7.1.1 and 7.1.2 can be extended without difficulty to the case $p \geq n$, the main difference being that no restriction on s is then necessary in Theorem 7.1.2; see Section 7.4.

In view of (7.1.1) no condition of p-regularity is needed for solutions. Theorems 7.1.1 and 7.1.2 are stated in [43] for the case $p = 2$, $n > 2$, and with the further differences: a_2, $\bar{a}_2$, b_1, b_2 are constants; the restriction $0 < s < (p-1)n/(n-p)$ is given as $1 < s < (p-1)n/(n-p)$; $\bar{a} \in L^{n/(1-\varepsilon)}(\Omega)$ rather than $\bar{a} \in L^n(\Omega)$.

Since the proofs are not entirely simple, we take special care to avoid undue conciseness.

---

[2]If $p = 1$ replace (6.1.2) by (6.1.3) as in Section 6.1.

Proof: Preliminaries. It is convenient to carry out the first part of the proofs of Theorems 7.1.1 and 7.1.2 in parallel. Also for simplicity, the calculations will be given in detail only for $p > 1$.

We assume initially that $R = 1$ (the general result then follows by rescaling, see (6.2.28) with $|\Omega| = \omega_n R^n$). The argument is similar to that in Lemmas 6.2.1 and 6.2.4, based on the study of auxiliary functions w.

We choose for test function

$$\varphi = \eta^p \, \psi(w), \qquad \eta = \eta(x), \qquad (7.1.5)$$

where η is a non-negative function in $C^1(\Omega)$ vanishing in a neighborhood of $\partial\Omega$, and ψ is a modified version of the function (6.2.1), depending on a non-zero real parameter $q \in \mathbb{R}$. Specifically, for $q \geq 1$ we take

$$\psi(t) = \frac{r^p}{q} \begin{cases} t^q, & \text{if} \quad 0 < t < m, \\ qm^{q-1}t - (q-1)m^q, & \text{if} \quad t \geq m, \end{cases}$$

while for $q < 1$ we use simply

$$\psi(t) = |r|^p \, t^q, \qquad t > 0,$$

where r is a real parameter given by the relation

$$q - 1 = p(r - 1). \qquad (7.1.6)$$

Note, in contrast to (6.2.1), the function ψ is now defined only for the range $t > 0$.

Step 1. For the proof of Theorem 7.1.1 *we restrict q to the range $q > 0$,* and take $w = u^+ + k$, $k = k(1)$. It can be supposed without loss of generality that $k > 0$, for otherwise replace k by $k' > 0$ and let $k' \to 0$ at the end of the proof. Since $\psi'(t)$ is uniformly bounded when $t \geq k$, both for $q \geq 1$ and $q < 1$, an obvious extension of Lemma 3.1.2 shows that $\varphi = \eta^p \psi(w)$ can serve as test function for (6.1.1), that is

$$\int_\Omega \langle A(x, u, Du), D\varphi \rangle \leq \int_\Omega [B(x, u, Du)]^+ \varphi. \qquad (7.1.7)$$

Clearly $\varphi = 0$, $D\varphi = \mathbf{0}$ a.e in the set where $w = k$. In the remaining set where $w > k$ we have $u > 0$, $u^+ = u$, that is

$$u = w - k, \qquad Du = Dw$$

and

$$D\varphi = \eta^p \psi'(w) Dw + p\eta^{p-1}\psi(w) D\eta.$$

As in the proof of Lemma 6.2.1 we define, for $q \geq 1$,

$$v(w) = \begin{cases} w^r, & \text{if} \quad k \leq w < m, \\ rm^{r-1}w - (r-1)m^r, & \text{if} \qquad w \geq m, \end{cases}$$

and simply $v = v(w) = w^r$ when $q < 1$. The inequality (7.1.7) can now be written explicitly, using a derivation parallel to that of (6.2.8). In particular, when $q \geq 1$ the estimates (6.2.7) continue to be valid, while when $q < 1$ they are replaced by the identities

$$\begin{aligned} \psi'(w)|Dw|^p &= q\,|Dv|^p, & \psi'(w)w^p &= q\,|r|^p\,v^p, \\ \psi(w)w^{p-1} &= |r|^p\,v^p, & \psi(w)|Dw|^{p-1} &= |r|\,v\,|Dv|^{p-1} \end{aligned} \tag{7.1.8}$$

(the absolute values for r are introduced for later purposes).

Therefore from (7.1.7), together with (6.1.2) and (7.1.1), we obtain after a short calculation

$$\begin{aligned} \int_\Omega |\eta Dv|^p \leq \mu \Bigg\{ &|r|\int_\Omega (b_1|\eta v| + pa_1|vD\eta|) \cdot |\eta Dv|^{p-1} \\ &+ |r|^p \int_\Omega p[\bar{a}_2 + (\bar{a}/k)^{p-1}] \cdot |\eta v|^{p-1}|vD\eta| \\ &+ |r|^p \int_\Omega [b_2 + (b/k)^{p-1}] \cdot |\eta v|^p \Bigg\} \\ &+ |r|^p \int_\Omega [a_2 + (a/k)^p] \cdot |\eta v|^p, \end{aligned} \tag{7.1.9}$$

where

$$\mu = \begin{cases} 1/|q|, & \text{if } q < 1, \\ 1/r, & \text{if } q \geq 1. \end{cases}$$

Step 1′. For the proof of Theorem 7.1.2 *we now restrict q to the range $q < 0$,* and take $w = u + k$ so $w \geq k$. Again it can be supposed without loss of generality that $k > 0$. As in the case $q < 1$ in Step 1 we define

$$\psi(t) = |r|^p\,t^q, \qquad t > 0,$$

in (7.1.5). As in Step 1, since $\psi'(t)$ is uniformly bounded when $t \geq k$ it is clear that $\varphi = \eta^q \psi(w)$ can serve as test function for the (reverse) inequality (7.1.4), that is

$$\int_\Omega \langle \boldsymbol{A}(x, u, Du), D\varphi \rangle \geq \int_\Omega [B(x, u, Du)]^+ \varphi. \tag{7.1.10}$$

Suppose $q \neq -(p-1)$, so $r \neq 0$, and define, as before in the case $q < 1$,

$$v = v(w) = w^r.$$

Then from (7.1.10), together with (6.1.2), (7.1.1), (7.1.2) and the identities (7.1.8), which holds equally when $q < 0$, we again obtain the inequality (7.1.9) *exactly in the form written*.

For the combined Steps 1 and 1' we have specifically, $r < 0$ when $q < -(p-1)$; $0 < r < 1/p'$ when $-(p-1) < q < 0$; and $r > 1/p'$ when $q > 0$. The anomalous case $q = -(p-1)$, $r = 0$, requires a separate treatment, see below.

Step 2. Define

$$z = \frac{\|\eta Dv\|_p}{\|\eta v\|_{p^*}}, \qquad y = \frac{\|\eta v\|_p}{\|\eta v\|_{p^*}}, \qquad \hat{y} = \frac{\|v D\eta\|_p}{\|\eta v\|_{p^*}}.$$

Then from the main identity (7.1.9) together with Sublemma 2 in the case $1 < p < n$, $s = p^*$, $\theta = 1$ (see Section 6.2), we get, after a short but straightforward calculation,

$$z^p \leq \mu|r|(\|b_1\|y^\varepsilon + pa_1\hat{y})z^{p-1} + (\mu+1)|r|^p c\, y^{p\varepsilon} + \mu|r|^p p\, \hat{c}\, \hat{y}, \qquad (7.1.11)$$

where[3]

$$c = \|a_2 + b_2\| + 2, \qquad \hat{c} = \|\bar{a}_2\| + 1.$$

From (7.1.11) and Lemma 6.2.2 we find next (with no attempt to give the sharpest estimate)

$$z \leq C_1 |r| (\mu+1)(y^\varepsilon + \hat{y}^{1/p} + \hat{y}),$$

where

$$C_1 = pa_1 + \|b_1\| + (pc)^{1/p} + (p2\hat{c})^{1/p}.$$

The Sobolev inequality implies[4] that $1 \leq S(z + \hat{y})$, so in turn

$$1 \leq SC_1 |r| (\mu+1)(y^\varepsilon + \hat{y}^{1/p} + \hat{y}) + S\hat{y}.$$

---

[3] If all the structural coefficients except a_1 vanish, then the right-hand side of inequality (7.1.11) retains only the term $a_1\tilde{y}z^{p-1}$, making the proof far simpler, and at the same time giving $k(R) = 0$, $C = C(p, n, s, a_1)$ in the theorems themselves. At a first reading of the proof it can be useful to consider only this case.
[4] We have $\|\eta v\|_{p*} \leq S\|D(\eta v)\|_p \leq S(\|\eta Dv\|_p + \|v D\eta\|_p)$.

By a double application of Young's inequality to rationalize the terms y^ε and $\hat{y}^{1/p}$, we then obtain

$$1 \le C\{|r|\,(\mu+1)\}^\nu(y+\hat{y}), \qquad \nu = \max\{1/\varepsilon,\, p\}, \qquad (7.1.12)$$

where C depends only on the parameters

$$p,\ n,\ \varepsilon; \qquad a_1 + \|b_1\| + \|a_2 + b_2\| + \|\bar{a}_2\|, \qquad (7.1.13)$$

while if $|r| < 1$ the term $|r|^\nu$ should be dropped from (7.1.12).

At this point it is no longer feasible to give precise estimates for the constants which appear in the calculations. Thus from here on, the letter C denotes generic constants depending only on the parameters (7.1.13).

Recalling the definitions of y and $\hat{y}$, application of (7.1.12) then gives

$$\|\eta v\|_{p*} \le C\{|r|(\mu+1)\}^\nu(\|\eta v\|_p + \|vD\eta\|_p). \qquad (7.1.14)$$

Still leaving aside the case $q = -(p-1)$, we now specify the function η more precisely. Let h, h' be such that

$$1 < h' < h < 3, \qquad (7.1.15)$$

and set $\eta \equiv 1$ in $B_{h'}$, $\eta \equiv 0$ in $\Omega \setminus B_h$, with $0 \le \eta \le 1$ and $\sup |D\eta| \le 2/(h-h')$ in $B_h \setminus B_{h'}$. (The last condition is obviously possible for $\eta \in C^1(\Omega)$.) Then from (7.1.14) there follows

$$\|v\|_{p*,B_{h'}} \le C\frac{\{(\mu+1)\,|r|\}^\nu}{h-h'}\|v\|_{p,B_h}. \qquad (7.1.16)$$

For $r \ne 0$ let us define

$$\Phi(r,h) = \left(\int_{B_h} w^r\right)^{1/r};$$

(this definition is meaningful since $w \ge k > 0$).

Suppose $r > 0$, $r \ne 1/p'$ and $\kappa = p^*/p = n/(n-p)$. Then (7.1.16) can be rewritten

$$\Phi(\kappa pr, h') \le \left(C\frac{\{(\mu+1)\,|r|\}^\nu}{h-h'}\right)^{1/r}\Phi(pr, h) \qquad (7.1.17)$$

(in case $r \ge 1$, that is $q \ge 1$, one first takes $m \to \infty$ so $v \to w^r$).

On the other hand, when $r < 0$ we have $\|w^r\|_{p,B_h} = \Phi(pr, h)^{-|r|}$ so there results instead (!)

$$\Phi(pr, h) \leq \left(C \frac{\{(\mu + 1)\, |r|\}^\nu}{h - h'} \right)^{1/|r|} \Phi(\kappa pr, h'). \tag{7.1.18}$$

These inequalities can be iterated as in the proof of Lemma 6.2.1.

Proof of Theorem 7.1.1. Here we consider parameter values $r > 1/p'$, $p > 1$, in which case $q > 0$ and Cases 1 and 2 above apply. Fix $s \in (p - 1, p]$ and take successively

$$r = r_j = \kappa^j (s/p); \qquad h = h_j = 1 + 2^{-j}, \qquad h' = h_{j+1},$$

$j = 0, 1, 2, \ldots$. Then from the definition of the parameter μ, the fact that $r \geq s/p$ and $q = p(r - 1) + 1$,

$$\mu = \max \left\{ \frac{1}{pr - (p - 1)}, \frac{1}{r} \right\} \leq \frac{1}{s - (p - 1)}, \qquad \frac{r^\nu}{h - h'} \leq 2 \, (2\kappa^\nu)^j.$$

Then by iteration of (7.1.17), as in the proof of Lemma 6.2.1, there results

$$\sup_{B_1} w \leq C\|w\|_{s,B_2};$$

where C depends on the parameters (7.1.13) and *also on s*. Theorem 7.1.1 for the case $p > 1$ now arises by taking $w = u^+ + k$ and then rescaling – see (6.2.28). Here s can be any value strictly greater than $p - 1$, by using Hölder's inequality.

The case $p = 1$ can be treated as in Lemma 6.2.4; we can omit the details. $\qquad \square$

Proof of Theorem 7.1.2. Here the argument is more delicate, as there are two regimes to consider: $r < 0$ and $0 < r < 1/p'$. In both ranges we have $q < 0$ by (7.1.6) so that Cases 1' and 2 apply.

Case A: $r < 0$. Let $s_1 \in (0, p)$ be fixed, and take successively

$$r = r_j = -\kappa^j (s_1/p); \qquad h = h_j = 2 + 2^{-j}, \qquad h' = h_{j+1},$$

$j = 0, 1, 2, \ldots$. Then

$$\mu = \frac{1}{p|r| + (p - 1)} \leq \frac{1}{p - 1}, \qquad \frac{|r|^\nu}{h - h'} \leq 2(2\kappa^\nu)^j.$$

Iteration of (7.1.18) then gives

$$\Phi(-s_1, 3) \le C^{1/s_1} \inf_{B_2} w; \tag{7.1.19}$$

here C depends on the parameters (7.1.13), while the exponent $1/s_1$ is included since the Moser iteration exponent, see (6.2.17), now includes the additional factor p/s_1.

Case B: $0 < r < 1/p'$, so $q < 1$. Fix $s \in (0, \kappa(p-1))$. For any integer $\ell = 0, 1, 2, \ldots$ define

$$s_2 = s_2(\ell) = \kappa^{-(\ell+1)} s, \qquad s_2 \in (0, p-1), \tag{7.1.20}$$

and choose successively

$$r = r_j = \kappa^j(s_2/p), \qquad h = h_j = 2 + 2^{-j}, \qquad h' = h_{j+1},$$

$j = 0, 1, 2, \ldots, \ell$. Then $0 < r < \kappa^{j-\ell}(s/\kappa p) \le s/\kappa p$, while

$$\mu = \frac{1}{p-1-pr} \le \frac{\kappa}{\kappa(p-1)-s}; \qquad \frac{|r|^\nu}{h-h'} \le 2 \cdot 2^j.$$

Then by a *finite iteration* of (7.1.17) from $j = 0$ to $j = \ell$ we obtain

$$\Phi(s, 2) \le C^{1/s_2} \Phi(s_2, 3), \tag{7.1.21}$$

where C depends on the parameters (7.1.13) and also on the fixed value s.

We claim that there exist constants $\sigma_0 > 0$ (depending only on the generic parameters (7.1.13)) and $\hat{C} = \hat{C}(n)$, such that for all $s_0 \le \sigma_0$,

$$\Phi(s_0, 3) \le \hat{C}^{1/s_0} \Phi(-s_0, 3). \tag{7.1.22}$$

Without loss of generality we can suppose $\sigma_0 < p - 1$.

Assume (7.1.22) for the moment. With s fixed as above in $(0, \kappa(p-1))$, now choose s_0 to be the unique value s_2 of the form (7.1.20) in the interval $[\sigma_0/\kappa, \sigma_0)$, thus fixing ℓ. Then (7.1.21) holds with $s_2 = s_0$; of course $s_0 < p - 1$. In turn (7.1.19) equally holds for the value $s_1 = s_0$. Therefore by (7.1.19) and (7.1.21) there would result

$$\|w\|_{s, B_2} = \Phi(s, 2) \le C^{1/s_0} \cdot C^{1/s_0} \cdot \hat{C}^{1/s_0} \inf_{B_2} w,$$

that is

$$\|w\|_{s, B_2} \le C \inf_{B_2} w,$$

where C depends on the generic parameters (7.1.13). This being shown, Theorem 7.1.2 is then proved by taking $w = u + k$ and rescaling.

It thus remains to prove the assertion (7.1.22). In fact, to obtain (7.1.22) it is remarkable that one can apply the previously omitted case $q = -(p-1)$.

Case C: $q = -(p-1)$. For this case the previously used test function ψ needs modification. We take simply $\psi(t) = t^{-(p-1)}$ and then define $v = v(w) = \log w$ (recall that $w = u + k \geq k > 0$). Then as in the derivation of (7.1.9), but with (7.1.8) replaced by

$$\psi'(w)|Dw|^p = -(p-1)|Dv|^p, \qquad \psi'(w)w^p = -(p-1),$$
$$\psi(w)w^{p-1} = 1, \qquad \psi(w)|Dw|^{p-1} = |Dv|^{p-1},$$

we get

$$(p-1)\int_\Omega |\eta Dv|^p \leq \int_\Omega (b_1\eta + pa_1|D\eta|) \cdot |\eta Dv|^{p-1}$$
$$+ \int_\Omega p[\bar{a}_2 + (\bar{a}/k)^{p-1}]\,\eta^{p-1}|D\eta| \qquad (7.1.23)$$
$$+ \int_\Omega [(p-1)(a_2 + (a/k)^p) + b_2 + (b/k)^{p-1}]\,\eta^p.$$

Let $x' \in B_3$ (recall $R = 1$, and denote by $B'_h = B_h(x')$ the ball of radius $h > 0$ centered at x' We now specify the test function η so that $\eta \equiv 1$ in B'_h and $\eta \equiv 0$ in $\Omega \setminus B'_{4h/3}$, with $0 \leq \eta \leq 1$ and $D\eta \leq 6/h$ in $B'_{4h/3} \setminus B'_h$. When $h < 3/4$ one has $B'_{4h/3} \subset\subset B'_1 \subset B_4$ since $x' \in B_3$, so η vanishes in a neighborhood of $\partial\Omega$, as required.

This being shown, the terms on the right side of (7.1.23) then have the following main estimates, in which C denotes as before different generic constants, depending on the parameters (7.1.13):

$$\int_\Omega pa_1|D\eta| \cdot |\eta Dv|^{p-1} \leq (6/h)pa_1\|1\|_{p,B'_h}\|\eta Dv\|_p^{p-1} \leq Ch^{(n-p)/p}\|\eta Dv\|_p^{p-1},$$

$$\int_\Omega b_1|\eta| \cdot |\eta Dv|^{p-1} \leq \|\eta\|_{np/(n-p)}\|b_1\|_n \|\eta Dv\|_p^{p-1} \leq Ch^{(n-p)/p}\|\eta Dv\|_p^{p-1},$$

$$\int_\Omega \bar{a}_2\eta^{p-1}|D\eta| \leq (6/h)\|\eta\|_{n(p-1)/(n-p+1)}^{p-1}\|\bar{a}_2\|_{n/(p-1)} \leq Ch^{n-p},$$

$$\int_\Omega (a_2 + b_2)\eta^p \leq \|\eta\|_{np/(n-p)}^p \|a_2 + b_2\|_{n/p} \leq Ch^{n-p}.$$

Therefore $\|\eta Dv\|_p^p \leq C[h^{(n-p)/p}\|\eta Dv\|_p^{p-1} + h^{n-p}]$.

Consequently, by Lemma 6.2.2

$$\|\eta Dv\|_p \leq C[h^{(n-p)/p} + h^{(n-p)/p}] = 2Ch^{(n-p)/p}. \tag{7.1.24}$$

Finally, by Hölder's inequality

$$\|Dv\|_{1,B_h'} \leq Ch^{n/p'}\|Dv\|_{p,B_h'} \leq Ch^{n-1}, \qquad 0 < h < 3/4. \tag{7.1.25}$$

We now use a remarkable theorem of John and Nirenberg (Appendix, Theorem 7.5.4) specifically in the case when the basic domain is the ball B_3. To apply this result requires that the condition

$$\int_{B_3 \cap B_h'} |Dv| \leq Kh^{n-1}, \qquad K = \text{constant}, \tag{7.1.26}$$

should be satisfied for *every* ball $B_h' = B_h(x')$, $h > 0$, with center x' in B_3.

To verify (7.1.26) there are two cases: $h < 3/4$ and $h \geq 3/4$. In the first, by (7.1.25) one immediately has

$$\int_{B_h'} |Dv| \leq Ch^{n-1}.$$

In the second case, we use (7.1.25) for the ball B_3 rather than B_h'. That is, since $B_{(4/3)\cdot 3} = B_4$ we can temporarily replace B_h' by B_3 in (7.1.25). Thus (7.1.25) yields

$$\int_{B_3} |Dv| \leq C \cdot 3^{n-1} \leq 4^{n-1}Ch^{n-1}$$

since $h \geq 3/4$. Therefore (7.1.26) holds for all $x' \in B_3$ and all $h > 0$, with $K = 4^{n-1}C$, where C is the constant in (7.1.25).

Hence by the John–Nirenberg theorem, in particular Corollary 7.5.6, we get

$$\int_{B_3} e^{s_0 v} \cdot \int_{B_3} e^{-s_0 v} \leq \tilde{C} = \tilde{C}(n)$$

for all $s_0 \leq \sigma_0$, where σ_0 is a (small) positive constant depending on n and the constant K, and so only on the generic parameters (7.1.13). But $v = \log w$ so that

$$\int_{B_3} w^{s_0} \cdot \int_{B_3} w^{-s_0} \leq \tilde{C},$$

that is,

$$\Phi(s_0, 3) \leq \tilde{C}^{1/s_0} \Phi(-s_0, 3)$$

(the constant $\tilde{C}^{1/s_0}$ can of course be written explicitly, and is surely a *very large* number). This proves (7.1.22) and completes the proof of Theorem 7.1.2. □

Corollary 7.1.3. *Let $u \in W_{loc}^{1,p}(\Omega)$, $p > 1$, be a non-negative solution of the inequality (7.1.4), where conditions (6.1.2), (7.1.1), (7.1.2) hold, with R_0^+ in place of R^+ and with $a = b = \bar{a} = 0$. Then either $u \equiv 0$ or $u > 0$ in Ω.*

This result has already been noted in Section 2.5.

7.2 The Harnack inequality

By combining Theorems 7.1.1 and 7.1.2 we obtain the full Harnack inequality.

Theorem 7.2.1 (Harnack Inequality). *Let $u \in W_{loc}^{1,p}(\Omega)$, $1 < p < n$, be a non-negative solution of the equation*

$$\mathrm{div}\,\boldsymbol{A}(x, u, Du) + B(x, u, Du) = 0. \tag{7.2.1}$$

Suppose that $\boldsymbol{A}$ and B satisfy (6.1.2) together with (7.1.1) and

$$|B(x, z, \boldsymbol{\xi})| \leq b_1 |\boldsymbol{\xi}|^{p-1} + b_2 z^{p-1} + b^{p-1}. \tag{7.2.2}$$

Then for any ball B_{4R} in Ω we have

$$\sup_{B_R} u \leq C[\inf_{B_{2R}} u + k(R)], \tag{7.2.3}$$

where C depends only on p, n, ε; a_1, $\|\bar{a}_2\|$, $R^\varepsilon\|b_1\|$, $R^{p\varepsilon}\|a_2 + b_2\|$, while $k(R)$ is given by (7.1.3).

Of course, all norms need be taken only over the ball B_{4R}.

Proof. Define $\bar{q} = (p-1)(n+p)/n$, so $p - 1 < \bar{q} < (p-1)n/(n-p)$. Hence both Theorems 7.1.1 and 7.1.2 apply for the value $q = \bar{q}$, that is

$$\sup_{B_R} u \leq C\left[R^{n/\bar{q}}\|u\|_{\bar{q}, B_{2R}} + k(R)\right] \leq C\left\{C\left[\inf_{B_{2R}} u + k(R)\right] + k(R)\right\}$$

$$= C^2 \inf_{B_{2R}} u + (C^2 + C)k(R) = 2\,C^2\left[\inf_{B_{2R}} u + k(R)\right],$$

as required. □

This result for the case $p = 2$ is given in [43, Theorem 8.20], though inadvertently the additive term $k(R)$ seems to have been omitted.

Theorem 7.2.2 (General Harnack Inequality). *Let u satisfy the hypotheses of Theorem 7.2.1. Then for any domain Ω' with compact closure in Ω we have*

$$\sup_{\Omega'} u \leq C^N \left[\inf_{\Omega'} u + Nk \right], \tag{7.2.4}$$

where $k = \|a\| + \|b\| + \|\bar{a}\|$ and N is the number of balls of (equal) radii $\delta/4$ needed to cover Ω', $\delta = \mathrm{dist}\,(\Omega', \partial\Omega)$. (We suppose $\delta \leq 4$ without loss of generality.)

Proof. Let $\mathscr{B}$ be a set of N open balls with centers in Ω' and equal radii $R = \delta/4$ which cover Ω'. Clearly there exist balls $B_I, B_S \in \mathscr{B}$ such that

$$\inf_{B_I} u \leq \inf_{\Omega'} u, \qquad \sup_{B_S} u \geq \sup_{\Omega'} u. \tag{7.2.5}$$

In turn, there exists a finite sequence of distinct balls, $B_{(1)}, B_{(2)}, \ldots, B_{(J)}$, $J \leq N$, in $\mathscr{B}$ such that[5]

$$B_{(1)} = B_I, \qquad B_{(J)} = B_S, \qquad B_{(i+1)} \cap B_{(i)} \neq \emptyset.$$

We claim that

$$\sup_{B_{(i)}} u \leq C^i L + \sum_{\ell=1}^{i} C^\ell k, \qquad L = \inf_{B_I} u, \tag{7.2.6}$$

for all $i = 1, \ldots, J$, where C is the constant of Theorem 7.2.1 taken for $R = 1$.

---

[5]This assertion has been considered obvious in earlier demonstrations (see, e.g., [43], Corollary 8.21, or [51], page 263, as well as even the original argument given by Harnack [45], page 62). Nevertheless, for completeness it seems worthwhile to indicate a proof:

Let x_I, x_S be respectively the centers of B_I and B_S, and Σ an oriented continuous curve in Ω' from x_I to x_S, existing since Ω' is open and connected. We set $B_{(1)} = B_I$. If $x_S \in B_{(1)}$, then we take $B_{(2)} = B_S$ and $J = 2$; or if $B_I = B_S$, then simply $J = 1$. Otherwise, if $x_S \notin B_{(1)}$, let $P_{(1)}$ be the last point where Σ intersects $\partial B_{(1)}$ (see Figure 1). We choose for $B_{(2)}$ any ball in $\mathscr{B}$ which contains $P_{(1)}$. If $x_S \in B_{(2)}$, we take $B_{(3)} = B_S$ and $J = 3$; or if $B_{(2)} = B_S$, then $J = 2$.

Continuing in this way, we thus obtain a finite sequence $B_{(1)}, B_{(2)}, \ldots, B_{(j)}$, of distinct balls in $\mathscr{B}$, such that $B_{(i+1)} \cap B_{(i)} \neq \emptyset$ and $x_S \in B_{(j)}$ for some $j \leq N$ (since $\mathscr{B}$ covers Ω'). If $B_{(j)} \neq B_S$ we take $B_{(j+1)} = B_S$, and $J = j+1 \leq N$, while if $B_{(j)} = B_S$, then simply $J = j$, and the assertion is proved.

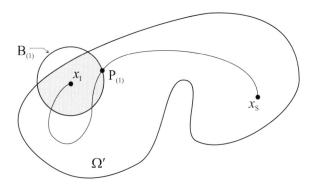

Figure 7.1.

Clearly (7.2.6) holds for $i = 1$; that is, by (7.2.2),

$$\sup_{B_{(1)}} u \le C \left[\inf_{B_{(1)}} u + k \right] = C(L + k)$$

since $4R \le \delta$. Thus suppose (7.2.6) is true for $i = j$; we shall show it true for $i = j + 1$, $j = 1, \ldots, N - 1$. Indeed by (7.2.2) again,

$$\sup_{B_{(j+1)}} u \le C \left[\inf_{B_{(j+1)}} u + k \right] \le C \left[\sup_{B_{(j)}} u + k \right]$$

since $B_{(j+1)} \cap B_{(j)} \ne \emptyset$. Then by (7.2.6) with $i = j$ we get

$$\sup_{B_{(j+1)}} u \le C \left(C^j L + \sum_{\ell=1}^{j} C^\ell k + k \right) = C^{j+1} L + \sum_{\ell=1}^{j+1} C^\ell k$$

as required.

The conclusion (7.2.4) is now an immediate consequence of (7.2.5) and (7.2.6) for $i = J$. $\qquad\square$

Both Theorems 7.2.1 and 7.2.2 also hold when $p \ge n$, see [92] and Section 7.4.

Corollary 7.2.3 (Liouville Theorem). *Let $u \in W^{1,p}_{\mathrm{loc}}(\mathbb{R}^n)$, $p > 1$, be a non-negative solution of the equation*

$$\mathrm{div}\, \mathbf{A}(x, u, Du) = 0$$

such that

$$\langle \boldsymbol{A}(x, u, \boldsymbol{\xi}), \boldsymbol{\xi} \rangle \geq |\boldsymbol{\xi}|^p, \qquad |\boldsymbol{A}(x, u, \boldsymbol{\xi})| \leq a_3 |\boldsymbol{\xi}|^{p-1}. \qquad (7.2.7)$$

Then $u \equiv$ Constant.

Proof. Let $M = \inf_{\mathbb{R}^n} u \ (\geq 0)$. Then $v = u - M$ is equally a solution of (7.2.7), with $\inf_{\mathbb{R}^n} v = 0$. For any $\varepsilon > 0$ there is some point $x_0 \in \mathbb{R}^n$ such that $v(x_0) \leq \varepsilon$ for any ball $B_R \subset \mathbb{R}^n$ with center at x_0. Hence by (7.2.3) we have $v(x) \leq C\varepsilon$.

Since ε is arbitrary there follows $v \leq 0$, that is $u \equiv M$. $\qquad \square$

When $\Omega = \mathbb{R}^n$ the proof technique used in Theorem 7.2.2 supplies an interesting asymptotic conclusion, which the authors have not previously seen.

To state the result, we first let $\mathscr{B}$ denote the set of balls of unit radius in $\mathbb{R}^n$ and put

$$k = \sup_{\mathscr{B}} \left\{ \|a\|_B + \|b\|_B + \|\bar{a}\|_B \right\}$$

and

$$m = \sup_{\mathscr{B}} \left\{ \sup_B a_1 + \|b_1\|_B + \|a_2 + b_2\|_B + \|\bar{a}_2\|_B \right\}.$$

Theorem 7.2.4. *Let u satisfy the hypotheses of Theorem 7.2.1, with $\Omega = \mathbb{R}^n$. Then $u \in L^\infty_{\mathrm{loc}}(\mathbb{R}^n)$ and*

$$u \leq [u(0) + kr] \, e^{Cr} \qquad \textit{a.e. in } \mathbb{R}^n, \qquad r = |x|, \qquad (7.2.8)$$

where C depends only on p, n, ε and m.

If the coefficients are all constants, than we can take $k = a + b + \bar{a}$ and $m = a_1 + b_1 + a_2 + b_2 + \bar{a}_2$, so that in this case necessarily u can have at most exponential growth at infinity. Of course, the choice of origin in (7.2.8) is arbitrary; the value $u(0)$ is meaningful, since in fact u must be continuous in view of the results of the following section.

7.3 Hölder continuity

The structural conditions of Theorem 7.2.1 also imply that solutions of the equation (7.2.1) are continuous. This remarkable result goes back to important work of De Giorgi [27].

In view of the general form of the result, the terms u^{p-1} and u^p in the basic structural conditions for A and B must be replaced by $|u|^{p-1}$ and $|u|^p$. Moreover, we shall add the strengthened Lebesgue space conditions that $\bar{a} \in L^{n/(1-\varepsilon)}(\Omega)$ and $\bar{a}_2 \in L^{n/(1-\varepsilon)(p-1)}(\Omega)$, rather than in $L^n(\Omega)$ and $L^{n/(p-1)}(\Omega)$ as previously required. Then solutions of (7.2.1) are in fact Hölder continuous. We state this famous result in precise form, and include the not entirely trivial proof.

Theorem 7.3.1 (Hölder Continuity). *Let u satisfy the hypotheses of Theorem 7.2.1, together with the modified structure conditions described above. Then u is Hölder continuous in Ω. Specifically, let x, $y \in \Omega$ and define*

$$D = \tfrac{1}{2}\operatorname{dist}(x, \partial\Omega).$$

Then if $|x - y| \leq D/4$ we have

$$|u(x) - u(y)| \leq [(M - m) + Ck] \left(\frac{4|x - y|}{D} \right)^\alpha, \qquad (7.3.1)$$

where

$$M = \sup_{B_D} u, \qquad m = \inf_{B_D} u, \qquad L = \sup_{B_D} |u|$$

and B_D denotes the ball of radius D centered at x. Moreover α depends only on n, p, ε, and the structural parameters a_1, $\|b_1\|$; C depends only on ε; while

$$k = \|a\| + \|b\| + \|\bar{a}\| + L \left[\|a_2\|^{1/p} + \|b_2\|^{1/(p-1)} + \|\bar{a}_2\|^{1/(p-1)} \right]. \quad (7.3.2)$$

(We suppose $D \leq 1$ without loss of generality.)

Note that $B_{2D} \subset \Omega$ so that by Theorem 7.1.1 the values M, m, L are well defined and finite. Also, we recall that by $\|a\|$, etc., we mean $\|a\|_{L^{n/(1-\varepsilon)}(\Omega)}$, etc.

From the proof it will appear that the key exponent α is essentially the reciprocal of the constant C in the Harnack inequality Theorem 7.2.1, evaluated for the arguments $R = 1$, $a_2 = b_2 = \bar{a}_2 = 0$, see (7.3.11). Of course even then α is very small. The constant C in (7.3.1) can be taken to be 4 when the structural coefficients are all constants, that is $\varepsilon = 1$.

By interchanging the roles of x and y it is clear that (7.3.1) actually holds with

$$D = \tfrac{1}{2}\max \{\operatorname{dist}(x, \partial\Omega), \operatorname{dist}(y, \partial\Omega)\}.$$

Proof. Step 1. Temporarily following the argument of the first part of the proof of Theorem 8 of [92] (see page 270), we let r be a radial variable having the range $(0, D)$, which throughout Step 1 we consider to be fixed. Define

$$\tilde{M} = M_r = \sup_{B_r} u, \qquad \tilde{m} = m_r = \inf_{B_r} u, \qquad M' = M_{r/4}, \qquad m' = m_{r/4}.$$

It follows that both functions

$$v = u - \tilde{m}, \qquad w = \tilde{M} - u$$

are non-negative in B_r. Obviously v satisfies the equation

$$\text{div } \boldsymbol{A}(x, v + \tilde{m}, Dv) + B(x, v + \tilde{m}, Dv) = 0,$$

while

$$|\boldsymbol{A}(x, v + \tilde{m}, Dv)| \leq a_1 |Dv|^{p-1} + \bar{a}_2 L^{p-1} + \bar{a}^{p-1},$$

with similar conditions for $\langle \boldsymbol{A}(x, v + \tilde{m}, Dv), Dv \rangle$ and $B(x, v + \tilde{m}, Dv)$. Thus we may apply Theorem 7.2.1 to v in the ball B_r, with the result

$$M' - \tilde{m} = \sup_{B_{r/4}} v \leq \lambda \left(\inf_{B_{r/4}} v + kr^\varepsilon \right) = \lambda(m' - \tilde{m} + kr^\varepsilon), \qquad (7.3.3)$$

where

$$\lambda = \lambda(p, n, \varepsilon, a_1, \|b_1\|) \ (> 1)$$

is the Harnack constant in Theorem 7.2.1 evaluated for the arguments $R = D$, $a_2 = b_2 = \bar{a}_2 = 0$, while k is given by (7.3.2). (Use $\bar{a} \in L^{n/(1-\varepsilon)}(\Omega)$, $\bar{a}_2 \in L^{n/(1-\varepsilon)(p-1)}(\Omega)$ in estimating the term $k(R)$ in Theorem 7.2.1.[6])

In the same way, w is non-negative in B_r and again by Theorem 7.2.1 we get

$$\tilde{M} - m' = \sup_{B_{r/4}} w \leq \lambda \left(\inf_{B_{r/4}} w + kr^\varepsilon \right) = \lambda(\tilde{M} - M' + kr^\varepsilon). \qquad (7.3.4)$$

---

[6]In view of the stronger spaces which are now assumed for $\bar{a}$ and $\bar{a}_2$, it is apparent that in the formulation of Theorems 7.1.1 and 7.2.1 one can take

$$\|\bar{a}\| \equiv \|\bar{a}\|_{n,B_{2R}} = c_n R^\varepsilon \|\bar{a}\|_{n/(1-\varepsilon)},$$

$$\|\bar{a}_2\| \equiv \|\bar{a}_2\|_{n/(p-1),B_{2R}} = c_n R^{(p-1)\varepsilon} \|\bar{a}_2\|_{n/(1-\varepsilon)(p-1)}.$$

Adding (7.3.3) to (7.3.4) and transposing terms then gives

$$M' - m' \leq \frac{\lambda - 1}{\lambda + 1}\left(\tilde{M} - \tilde{m}\right) + \frac{2\lambda}{\lambda + 1} k r^{\varepsilon}.$$

Letting $\omega(r)$ denote the oscillation of u in B_r, this can be rewritten

$$\omega(r/4) \leq \vartheta\omega(r) + \ell k r^{\varepsilon}, \tag{7.3.5}$$

where

$$\vartheta = (\lambda - 1)/(\lambda + 1), \qquad \ell = 2\lambda/(\lambda + 1).$$

This is the key inequality for proving (7.3.1).

Step 2. Using ν (≥ 1) successive iterations of (7.3.5) to smaller values of r, there results for $\nu = 1, 2, \ldots$

$$\omega\left(4^{-\nu}r\right) \leq \vartheta^{\nu}\omega(r) + \ell k r^{\varepsilon}\left\{\vartheta^{\nu-1} + 4^{-\varepsilon}\vartheta^{\nu-2} + 4^{-(\nu-1)\varepsilon}\right\}$$

$$\leq \vartheta^{\nu}\omega(r) + \ell k r^{\varepsilon}\sum_{j=0}^{\nu-1} 4^{-\varepsilon j} \tag{7.3.6}$$

$$\leq \vartheta^{\nu}\omega(D) + C' k r^{\varepsilon}, \qquad C' = \frac{2}{1 - 4^{-\varepsilon}},$$

since $r \leq D$ and $\ell \leq 2$.

Let $\beta > 1$ be a constant to be determined later. We take for the radius variable r in (7.3.6) the particular ν-dependent choice

$$r_{\nu} = 4^{(1-\nu)(\beta-1)} D,$$

which is allowable since $\nu \geq 1$. Now define $t = t_{\nu} = 4^{-\nu} r_{\nu}$ and observe by direct arithmetic that

$$r_{\nu} = (4t/D)^{1/\beta'} D. \tag{7.3.7}$$

Hence

$$\omega(t) \leq \vartheta^{\nu}\omega(D) + C' k \left(4t/D\right)^{\varepsilon/\beta'}, \tag{7.3.8}$$

using here the condition that $D \leq 1$.

Next, putting $d = 2/(1 + \lambda)$,

$$\log \vartheta = \log(1 - d) \leq -d + d^2 = -\frac{2}{\lambda} \cdot \frac{\lambda(\lambda - 1)}{(\lambda + 1)^2} \leq -36/25 \, \lambda,$$

it being assumed without loss of generality that $\lambda \geq 9$. But then

$$\vartheta^\nu = e^{\nu \log \vartheta} \leq 4^{-36\nu/25\lambda \log 4} \leq 4^{-\nu/\lambda}. \tag{7.3.9}$$

At the same time, by (7.3.7),

$$4^{-\nu} = \frac{t}{r_\nu} = \frac{t}{(4t/D)^{1/\beta'}D} = \frac{1}{4}\left(\frac{4t}{D}\right)^{1/\beta}.$$

Thus by (7.3.8) and (7.3.9) we obtain

$$w(t) \leq 4^{-1/\lambda}w(D)\left(\frac{4t}{D}\right)^{1/\lambda\beta} + C'k\left(\frac{4t}{D}\right)^{\varepsilon/\beta'}, \tag{7.3.10}$$

this being valid for the particular iterates $t = t_\nu$, $\nu = 1, 2, \ldots$.
Now choose $\beta = 1 + 1/\lambda\varepsilon$. Then[7]

$$\frac{1}{\beta\lambda} = \frac{\varepsilon}{\beta'} = \frac{\varepsilon}{1 + \lambda\varepsilon} \geq \frac{\varepsilon}{\lambda + 1} = \alpha \tag{7.3.11}$$

(defining α). Then (7.3.10) can be rewritten

$$w(t) \leq [4^{-1/\lambda}w(D) + C'k]\left(\frac{4t}{D}\right)^\alpha,$$

again valid for $t = t_\nu$.

The restriction of t to the iterate values t_ν can be removed by observing that, for $t_{\nu+1} < t \leq t_\nu$,

$$w(t) \leq w(t_\nu) \leq [4^{-1/\lambda}w(D) + C'k]\left(\frac{4}{D} \cdot \frac{t_\nu}{t_{\nu+1}} t\right)^\alpha.$$

Here

$$\frac{t_\nu}{t_{\nu+1}} = 4\frac{r_\nu}{r_{\nu+1}} = 4 \cdot 4^{\beta-1} = 4^\beta, \qquad \left(\frac{t_\nu}{t_{\nu+1}}\right)^\alpha = 4^{\beta\alpha} \leq 4^{1/\lambda}.$$

Hence, noting that $w(D) = M - m$, we obtain

$$w(t) \leq (M - m + Ck)(4t/D)^\alpha, \qquad C = \frac{3}{1 - 4^{-\varepsilon}}, \tag{7.3.12}$$

valid for all $t \leq t_1 = D/4$.

---

[7]In the special case $k = 0$ (as in the original papers of De Giorgi [27] and Moser [62]) we can take $\beta = 1$ so that $r_\nu = D$ and $\alpha = 1/\lambda$.

Step 3. For any $x \in \Omega$ and $0 < h \leq D/4$, define

$$M_h = \sup_{B_h(x)} u, \qquad m_h = \inf_{B_h(x)} u, \qquad u_h(x) = \frac{1}{|B_h(x)|} \int_{B_h(x)} u.$$

Clearly $m_h \leq u_h(x) \leq M_h$. Also, since the sequences m_h, M_h are monotone in h as $h \to 0$, they converge to limits. On the other hand, by (7.3.10) we have $M_h - m_h = \omega(h) \to 0$ as $h \to 0$, so that the limits must be the same. Therefore $u_h(x)$ converges to this limit, which we temporarily call $\tilde{u}(x)$.

By the Lebesgue set theorem, if u is *any* representative of u, we also have $u_h(x) \to u(x)$ at a.a. $x \in \Omega$. Hence $\tilde{u} = u$ a.e. in Ω, that is $\tilde{u}$ is a representative of u.

Step 4. We show that u is continuous, that is, the representative $\tilde{u}$ is a continuous function.[8]

Let x, y in Ω, with $4|x - y|/D < 1$. We put $t = |x - y|$ so that both x and y are in $B_t(x)$. It follows that, for $0 < h < 3D/4$,

$$|u_h(x) - u_h(y)| \leq \sup_{B_h(x)} u - \inf_{B_{t+h}(x)} u = \omega(t + h).$$

If h is sufficiently small, then $t + h \leq D/4$. Thus by (7.3.12)

$$|u_h(x) - u_h(y)| \leq (M - m + Ck)\left[4(t + h)/D\right]^\alpha. \qquad (7.3.13)$$

Letting $h \to 0$ then yields (7.3.1) with u replaced by $\tilde{u}$. That is, $\tilde{u}$ is continuous in Ω, and, by dropping the tilde, we get (7.3.1) as written. □

If only $\bar{a} \in L^n(\Omega)$ and $\bar{a}_2 \in L^{n/p}(\Omega)$, then by a modification of the above proof it is still possible to show that u is continuous, though no longer Hölder continuous.

7.4 The case $p \geq n$

By appropriately reducing the exponent $p = n$ using the Hölder inequality, it is not hard to see that one can extend the results of the previous sections to the exponent range $p \leq n$. At the same time, with the help of Morrey's theorem (Theorem 7.5.7) asserting the Hölder continuity of functions in $W^{1,p}(\Omega)$, $p > n$, it is possible to obtain these results also for the remaining range $p > n$, in fact by a direct and immediate route.

---

[8]This step has been omitted in earlier proofs, along with the construction of $\tilde{u}(x)$.

Theorem 7.4.1. *Let the hypotheses of Theorem 7.1.2 be satisfied, with the exception that $p > n$ and that the coefficients a, b, etc., are in the respective Lebesgue spaces:*

$$a,\ \bar{a},\ b_1 \in L^p(\Omega); \quad b \in L^{p-1}(\Omega); \quad a_2,\ b_2 \in L^1(\Omega); \quad \bar{a}_2 \in L^{p'}(\Omega). \quad (7.4.1)$$

Let $u \in W^{1,p}_{loc}(\Omega)$ be a positive solution of (7.1.4) in Ω. If $B_{4R} \subset \Omega$, then the following conclusions hold:

(i)
$$\sup_{B_R} u \le C \left(\inf_{B_R} u + k \right),$$

(ii) $|u(x) - u(y)| \le C \left(\sup_{B_{3R}} u + k \right) |x - y|^{1-n/p}$ *for all x, $y \in B_R$,*

where
$$C = C(n, p, R; a_1, \|b_1\|, \|a_2 + b_2\|, \|\bar{a}_2\|),$$
$$k = R^{(n-p)/p} \left[\|a\| + \|\bar{a}\| + R^{p'} \|b\| \right],$$

the norms of the coefficients being taken in the respective Lebesgue spaces (7.4.1).

Proof. Suppose first that $R = 1$. Observe that the key inequality (7.1.23) in Case (C) of the proof of Theorem 7.1.2 is equally valid when $p > n$. The following main estimates similarly remain true, *provided the parameter n appearing in these estimates is replaced by p and of course using the conditions* (7.4.1). Consequently in place of (7.1.24) we reach the conclusion

$$\|Dv\|_{p,B_3} \le C, \qquad (p > n)$$

since $B_4 \subset \Omega$.

Theorem 7.5.7 can therefore be applied, with the result

$$|v(x) - v(y)| \le C|x - y|^{1-n/p} \le C$$

when x, $y \in B_1$.

But $v = \log w = \log(u + k)$, so

$$u(x) \le \{u(y) + k\}e^C,$$

and (i) is an immediate consequence.

Moreover, since $\|Dv\|_{p,B_3} = \|Dw/w\|_{p,B_3}$ we have

$$\|Dw\|_{p,B_3} \le \sup_{B_3} w \|Dv\|_{p,B_3} \le C \sup_{B_3} w.$$

Therefore, again by Morrey's theorem,

$$|w(x) - w(y)| \le C|x - y|^{1-n/p} \sup_{B_3} w,$$

from which (ii) follows at once. □

The Harnack inequality (i) is an exact counterpart of (7.2.3), but interestingly it requires only that u be a solution of the inequality (7.1.4) rather than equation (7.2.1). Similarly, the Hölder inequality (ii) corresponds to (7.3.1), but with a far better exponent. At the same time, one should note that the constants C in (i) and (ii) becomes infinite as p approaches n.

Finally, the reader can easily convince himself that global bounds for solutions, of the kind developed in Chapter 6, *cannot* be obtained with the aid of the Harnack theorems or the Hölder continuity results of this chapter. That is, the results of Chapter 6 retain their validity irrespective of the theorems of this chapter.

7.5 Appendix. The John–Nirenberg theorem

This well-known result is crucial in the proof of the Harnack inequality in Section 7.1. For completeness we include a concise proof based only on Hölder's inequality and integration by parts, following an idea of Trudinger [110]. We begin with an important result concerning the Morrey transform of a function $f \in L^1(\Omega)$, this being the key to the main Theorem 7.5.4.

Proposition 7.5.1. *Suppose $n \ge 2$. Let $f \in L^1(\Omega)$ be such that*

$$f \ge 0 \quad in \ \Omega, \qquad \int_{\Omega \cap B'_h} f \le h^{n-1} \tag{7.5.1}$$

for all balls $B'_h = B_h(x')$ with radius $h > 0$ and center $x' \in \Omega$. Then there are positive constants a_1 and a_2, depending only on n, such that

$$\int_\Omega \exp\left(\frac{\mathscr{F}}{a_1}\right) \le a_2 d^n,$$

where $d = \operatorname{diam}(\Omega)$ and

$$\mathscr{F}(x) = \int_\Omega |x - y|^{1-n} f(y) dy$$

is the Morrey transform of f.

One can take $a_1 = 5n$ and $a_2 = \omega_n$. Further results relating to Proposition 7.5.1, but beyond the scope of the present work, can be found in Section 2.9 of [117] and in [98]. For the proof of the proposition two preliminary lemmas are required.

Lemma 7.5.2. *Let $f \in L^1(\Omega)$ and define* (Riesz potential)

$$\mathscr{F}_\theta(x) = \int_\Omega |x - y|^{n(\theta-1)} f(y) dy, \qquad \theta \in (0, 1].$$

Then

$$\|\mathscr{F}_\theta\|_1 \le \omega_n^{1-\theta} |\Omega|^\theta \|f\|_1 / \theta.$$

Proof. Let $\mathscr{S}(x) = |x|^{n(\theta-1)}$. Then $\mathscr{S} \in L^1(\Omega)$, with

$$\|\mathscr{S}\|_1 \le \omega_n^{1-\theta} |\Omega|^\theta / \theta.$$

Indeed, let $\mathscr{R}$ be the *effective radius* of Ω, that is $|\Omega| = \omega_n \mathscr{R}^n$. Then, since $n(\theta - 1) < 0$ one sees by geometric comparison that (!)

$$\int_\Omega \mathscr{S}(x) dx \le \int_{B_\mathscr{R}} \mathscr{S}(x') dx' = n\omega_n \int_0^\mathscr{R} \varrho^{n\theta-1} d\varrho = \frac{\omega_n}{\theta} \mathscr{R}^{n\theta},$$

as asserted. Now by Fubini's theorem

$$\int_\Omega \mathscr{F}_\theta(x) dx = \int_\Omega \left\{ \int_\Omega |x - y|^{n(\theta-1)} dx \right\} f(y) dy \le \frac{1}{\theta} \omega_n^{1-\theta} |\Omega|^\theta \int_\Omega f(y) dy,$$

since the center at y rather than at 0 leaves the estimates unchanged. $\square$

That $\mathscr{F}$ is well defined and in $L^1(\Omega)$ follows immediately from Lemma 7.5.2, since $\mathscr{F} = \mathscr{F}_{1/n}$. The second lemma is the ultimate key to the John–Nirenberg theorem.

Lemma 7.5.3. *Let $f \in L^1(\Omega)$ satisfy* (7.5.1). *For $\theta = (1 + \lambda)/n$, $0 < \lambda \le n - 1$, the function $\mathscr{F}_\theta$ defined in Lemma 7.5.2 is such that*

$$\mathscr{F}_\theta(x) \le \frac{n-1}{\lambda} d^\lambda \qquad \text{in } \Omega.$$

Before proving Lemma 7.5.3 it is useful to recall the following general integration by parts theorem.

Suppose that ϕ, ψ are absolutely continuous functions[9] on the bounded interval $[a, b]$. Then ϕ, ψ are differentiable a.e. on (a, b) with ϕ', $\psi' \in L^1(a, b)$, and

$$\phi(b)\psi(b) - \phi(a)\psi(a) = \int_a^b \{\phi(s)\psi'(s) + \psi(s)\phi'(s)\}ds. \qquad (7.5.2)$$

The proof is immediate. That is, since ϕ, ψ are absolutely continuous, also $\phi\psi$ is absolutely continuous. Thus $\phi\psi$ is differentiable a.e. on (a, b) and (see [89]) equals the indefinite integral of its derivative. But by direct evaluation $(\phi\psi)' = \phi\psi' + \psi\phi'$ a.e. in (a, b) and the conclusion follows at once. $\qquad \square$

Proof of Lemma 7.5.3. Without loss of generality we may suppose that the domain of f is extended to all $\mathbb{R}^n$ by setting $f \equiv 0$ outside Ω.

Let $\varepsilon \in (0, d)$ and $x \in \Omega$ be fixed, and define

$$\phi(t) = \int_{B_t(x)\setminus B_\varepsilon(x)} f(y)dy, \qquad t \in (\varepsilon, d).$$

Writing ϕ as an iterated integral in spherical polar coordinates r, ω,

$$\phi(t) = \int_\varepsilon^t r^{n-1} \int_{S^{n-1}} f(y(r, \omega)) \, d\omega \, dr,$$

shows that ϕ is absolutely continuous in the interval $[\varepsilon, d]$, see [89], Proposition 4.13, with derivative $\phi' \in L^1(\varepsilon, d)$. Also define

$$F(t) = \int_{B_t(x)\setminus B_\varepsilon(x)} |x - y|^{n(\theta-1)} f(y)dy.$$

Similarly, F is absolutely continuous in (ε, d), and one derives also

$$F'(t) = t^{n(\theta-1)}\phi'(t) \qquad \text{a.e. in } (\varepsilon, d).$$

Hence

$$F(d) = \int_\varepsilon^d F'(s) \, ds = \int_\varepsilon^d s^{n(\theta-1)}\phi'(s) \, ds$$

$$= d^{n(\theta-1)}\phi(d) + n(1 - \theta) \int_\varepsilon^d s^{n(\theta-1)-1}\phi(s) \, ds$$

by integration by parts, with $(a, b) = (\varepsilon, d)$ and $\psi(t) = t^{n(\theta-1)}$.

---

[9]For a bird's eye view of the properties of absolutely continuous functions, see [89], Section 5.4. The conclusion (7.5.2) does not hold if the hypothesis is weakened to the simple assertion that ϕ, ψ are continuous and differentiable a.e. in $[a, b]$ with ϕ', $\psi' \in L^1(a, b)$.

By (7.5.1) and the condition $f \equiv 0$ in $\mathbb{R}^n \setminus \Omega$, we have

$$h^{n(\theta-1)}\phi(h) \le h^{n(\theta-1)} \int_{\Omega \cap B_h(x)} f(y)dy \le h^{n\theta-1} = h^\lambda$$

since $\theta = (1+\lambda)/n$. Hence

$$F(d) \le \left[1 + \frac{n(1-\theta)}{\lambda}\right] d^\lambda = \frac{n-1}{\lambda} d^\lambda.$$

Using the fact that $\Omega \subset B_d(x)$, it follows that

$$F(d) = \int_{\Omega \setminus B_\varepsilon(x)} |x-y|^{n(\theta-1)} f(y)dy.$$

Lemma 7.5.3 is now a consequence of the monotone convergence theorem applied to the integral $F(d)$ as $\varepsilon \to 0$. $\qquad\square$

It is interesting that when (7.5.1) holds the integral $\mathscr{F}_\theta(x)$ is convergent for *all* x in Ω.

Proof of Proposition 7.5.1. Let $\lambda \in (0,1]$. We have

$$\frac{1}{n} - 1 = \left(\frac{\lambda}{n} - 1\right)\lambda + \left(\frac{1+\lambda}{n} - 1\right)(1-\lambda).$$

Then for $y \in \Omega$,

$$|x-y|^{1-n} f(y) \le \left\{|x-y|^{n[\lambda/n-1]} f(y)\right\}^\lambda \cdot \left\{|x-y|^{n[(1+\lambda)/n-1]} f(y)\right\}^{1-\lambda}.$$

Therefore, from Hölder's inequality with exponents $1/\lambda$ and $1/(1-\lambda)$,

$$\mathscr{F}(x) \le [\mathscr{F}_{\lambda/n}(x)]^\lambda [\mathscr{F}_{(1+\lambda)/n}(x)]^{1-\lambda} \le [\mathscr{F}_{\lambda/n}(x)]^\lambda \left(\frac{n-1}{\lambda} d^\lambda\right)^{1-\lambda}.$$

In turn, using Lemma 7.5.2,

$$\int_\Omega [\mathscr{F}(x)]^{1/\lambda} dx \le \frac{n}{\lambda} \omega_n^{1-\lambda/n} |\Omega|^{\lambda/n} \|f\|_1 \cdot \left(\frac{n-1}{\lambda} d^\lambda\right)^{1/\lambda-1}$$

$$\le \frac{n}{n-1} \left(\frac{n-1}{\lambda}\right)^{1/\lambda} \omega_n d^n$$

since $\|f\| \le d^{n-1}$ by (7.5.1) while obviously $|\Omega| \le \frac{1}{2}\omega_n d^n$.

Consequently, replacing $1/\lambda$ successively by $k = 1, 2, \ldots, N$, we get

$$\int_\Omega \sum_{k=1}^N \frac{1}{k!} \left[\frac{\mathscr{F}(x)}{a_1} \right]^k dx \le \frac{n}{n-1} \omega_n d^n \sum_{k=1}^N \frac{k^k}{k!} \left(\frac{n-1}{a_1} \right)^k$$

for any $a_1 > 0$. The ratio of successive terms in the series on the right-hand side is $(1 + 1/k)^k (n-1)/a_1 \le e(n-1)/a_1$. Hence, taking $a_1 = 5n$, the right-hand series is dominated by

$$\frac{n-1}{5n} \sum_{k=0}^{N-1} \left[\frac{(n-1)e}{5n} \right]^k \le \frac{n-1}{5n - (n-1)e}.$$

Finally, by the monotone convergence theorem,

$$\int_\Omega \exp\left(\mathscr{F}(x)/a_1 \right) dx \le \left[\frac{1}{2} + \frac{n}{5n - (n-1)e} \right] \omega_n d^n < \omega_n d^n,$$

as required. □

We can now prove the following version of the John–Nirenberg theorem.[10]

Theorem 7.5.4. *Let Ω be a bounded convex domain, and assume that $v \in W^{1,1}(\Omega)$, with*

$$\int_{\Omega \cap B'_h} |Dv| dx \le K h^{n-1}$$

for all balls $B'_h = B_h(x')$ with radius $h > 0$ and center $x' \in \Omega$.
  Then

$$\int_\Omega \exp\left(\frac{\sigma}{K} |v(x) - v_\Omega| \right) dx \le \omega_n d^n, \qquad v_\Omega = \frac{1}{|\Omega|} \int_\Omega v(x) dx$$

for all $\sigma \le |\Omega|/5d^n$, where $d = \mathrm{diam}(\Omega)$.

The proof of Theorem 7.5.4 is a consequence of Proposition 7.5.1 together with the following lemma, due originally to Morrey.

Lemma 7.5.5. *Let Ω be a bounded convex domain and $f = |Dv| \in L^1(\Omega)$. Then for a.a. $x \in \Omega$,*

$$|v(x) - v_\Omega| \le \frac{d^n}{n|\Omega|} \cdot \mathscr{F}(x).$$

For proof we refer the reader to [43, Lemma 7.16].

---

[10] The result as stated here is slightly weaker than the original theorem in [48]; it can be obtained by the original theorem together with Poincaré's inequality. The present proof however is both simpler and more concise.

Proof of Theorem 7.5.4. After a simple rescaling of v, the result follows by combining Proposition 7.5.1 and Lemma 7.5.5.   □

Corollary 7.5.6. *Under the hypotheses of Theorem 7.5.4 there holds*

$$\int_\Omega e^{\sigma v/K} \cdot \int_\Omega e^{-\sigma v/K} \leq 2\omega_n \, d^{2n}.$$

This is a consequence of the relations $v \leq |v - v_\Omega| + v_\Omega$ and $-v \leq |v - v_\Omega| - v_\Omega$.

It is interesting to observe that by Jensen's inequality

$$\int_\Omega e^{\sigma v/K} \cdot \int_\Omega e^{-\sigma v/K} \geq |\Omega|^2.$$

The John–Nirenberg theorem is closely related to a famous theorem of Morrey concerning the Hölder continuity of functions whose gradient is in $L^p(\Omega)$, $p > n$.

Theorem 7.5.7 (Morrey's theorem). *Let $u \in W^{1,p}(\Omega)$, $p > n$. Then u is locally Hölder continuous in Ω with Hölder exponent $1 - n/p$. Moreover, if Ω is convex, then also*

$$|u(x) - u(y)| \leq 2\, C(n,p)|\Omega|^{1/n - 1/p - 1} d^n \|Du\|_p, \qquad x,\, y \in \Omega, \quad (7.5.3)$$

where

$$C(n,p) = n^{-1/p} \left(\frac{p-1}{p-n} \right)^{1/p'} \omega_n^{1-1/n}.$$

Proof. First, by essentially the same proof as that for Lemma 7.5.2 we have

$$\|\mathscr{F}\|_\infty \leq n C(n,p)|\Omega|^{1/n - 1/p}\|Du\|_p.$$

Suppose now that Ω is convex. Then by Lemma 7.5.5 we have (with $f = |Du|$)

$$|u(x) - u_\Omega| \leq \frac{d^n}{n|\Omega|}\mathscr{F}(x) \leq C(n,p)|\Omega|^{1/n - 1/p - 1} d^n \|Du\|_p.$$

In turn, for $x,\, y \in \Omega$,

$$|u(x) - u(y)| \leq |u(x) - u_\Omega| + |u(y) - u_\Omega|,$$

so (7.5.3) follows at once.

Suppose $\mathrm{dist}(x, \partial\Omega) = \delta$ and let y be such that $|x - y| = h < \delta$. Obviously x, $y \in B_h(x)$. Taking $\Omega = B_h(x)$, so $d = 2h$, then from (7.5.3) follows

$$|u(x) - u(y)| \leq 2^{n+1} \left(\frac{1}{n\omega_n}\right)^{1/p} \left(\frac{p-1}{p-n}\right)^{1/p'} \|Du\|_p \cdot |x - y|^{1-n/p} \quad (7.5.4)$$

whenever $|x - y| < \delta$, proving that u is Hölder continuous. $\qquad\square$

Morrey's inequality, Theorem 3.9.3, is an easy consequence of (7.5.4). Indeed, if $|\Omega| = 1$, then every point of Ω is at most a distance $\omega_n^{-1/n}$ from $\partial\Omega$. Then by (7.5.4), with $y \in \partial\Omega$ so $|x - y| \leq \omega_n^{-1/n}$, we get

$$|u(x)| \leq \frac{2^{n+1}}{\omega_n^{1/n} n^{1/p}} \left(\frac{p-1}{p-n}\right)^{1/p'} \|Du\|_p, \quad (7.5.5)$$

where we have used the fact that $u(y) = 0$ because $u = 0$ on $\partial\Omega$.

Notes

Theorems 7.1.1 and 7.2.1 are essentially due to Serrin [92], Theorems 1 and 5. They are based ultimately on the Moser iteration technique. Theorem 7.1.2, as an intermediate step between Theorems 7.1.1 and 7.2.1, was first explicitly stated by Trudinger [109]: its great usefulness, as pointed out by Trudinger, lies in the fact that it applies to the differential inequality (7.1.4), rather than requiring the full differential equation (7.2.1) for its validity. The proofs in Section 7.1 are due to Serrin [92], with important modifications for clarity of presentation.

The results in Sections 7.2–7.4 are standard, but the statements and proofs are in many respects new; see especially Theorem 7.4.1.

Harnack inequalities for domains $\Omega \subset \mathbb{R}^2$ have been obtained by Pucci and Serrin [80] and [83, Section 5.5]. While the restriction to $\mathbb{R}^2$ is a drawback, on the other hand the operators and nonlinearities studied in this work are more general than in earlier literature, for example applying to the mean curvature equation even without bounds either on the solution or its gradient.

For mean curvature-type equations Trudinger [111] has given a Harnack inequality for *bounded* solutions in n dimensions, with the constant in the Harnack principle depending on the bound.

Problems

7.1 Supply the details for the proof of (7.1.11).

7.2 Using the proof method of Lemma 6.2.4, prove the case $p = 1$ of Theorem 7.1.1.

7.3 Prove Corollary 7.1.3.

7.4 If $\bar{a} \in L^n(\Omega)$ and $\bar{a}_2 \in L^{n/p}(\Omega)$ in Theorem 7.3.1, then show that u is continuous. Produce an example in which u is no longer Hölder continuous.

7.5 Prove Lemma 7.5.5, adapting the proof of [43, Lemma 7.16].

Chapter 8

Applications

8.1 Cauchy–Liouville Theorems

A Cauchy–Liouville type theorem is a statement that under appropriate circumstances an entire solution (a solution defined over $\mathbb{R}^n$) of an elliptic equation must be constant.[1] For the Laplace equation in particular, it is enough that a solution u should be bounded, or even, at a minimum, that $u(x) = o(|x|)$ as $|x| \to \infty$. For quasilinear equations, and even for semilinear equations of the form

$$\Delta u + B(u, Du) = 0, \qquad x \in \mathbb{R}^n, \tag{8.1.1}$$

the same question is more delicate than might at first be expected, since a number of different kinds of behavior can be seen even for relatively simple examples.

Consider first the simple Poisson equation

(I) $$\Delta u = f(u), \qquad u \in C^2(\mathbb{R}^n),$$

in which $f(u)$ is a non-decreasing function. If $u(x) = o(|x|)$ as $|x| \to \infty$, then $u \equiv \text{constant}$. For the equation

(II) $$\Delta u = \left(|Du|^2 - 1\right)^2 u, \qquad u \in C^2(\mathbb{R}^n),$$

---

[1]Frequently called Liouville theorems in the literature. For a discussion of the relative contributions of Cauchy and Liouville, see reference [101].

the same result holds, and indeed, more precisely $u \equiv 0$. On the other hand, in contrast to the Laplace equation, a one sided bound on u is not enough to make $u \equiv$ constant, since one can check that both

$$u(x) = \sqrt{1 + x_1^2}, \qquad u(x) = -\sqrt{1 + x_1^2}$$

are solutions. In a third case

(III) $\qquad\qquad\qquad \Delta u = |Du|^2, \qquad u \in C^1(\mathbb{R}^n),$

the *only* entire solutions are constants, without placing any bound on the solution itself. Even more the equation

(IV) $\qquad\qquad\qquad \Delta u = |Du|^2 + a, \qquad a = \text{constant} \neq 0,$

has no bounded entire solutions whatsoever.

Case (III) is proved by making the substitution $v = e^{-u}$, whence $\Delta v = 0$, $v > 0$, so that v and hence u must be constants.

Cases (I), (II) and (IV) rely on the following subtle lemma, which we state in greater generality than initially needed.

Theorem 8.1.1. *Consider the quasilinear equation*

$$a_{ij}(x, u, Du)\partial^2_{x_i x_j} u + B(x, u, Du) = 0, \qquad x \in \mathbb{R}^n, \qquad (8.1.2)$$

in which $[a_{ij}(x, z, \boldsymbol{\xi})]$ *is an* $n \times n$ *non-negative definite matrix, uniformly bounded in* $\mathbb{R}^n \times \mathbb{R} \times \boldsymbol{B}_\delta$ *for some* $\delta > 0$, *where* $\boldsymbol{B}_\delta$ *denotes the* δ-*ball of* $\mathbb{R}^n$. *Assume also that for* $x \in \mathbb{R}^n$

$$-B(x, z, \boldsymbol{\xi}) \begin{cases} \geq f(z) - g(|\boldsymbol{\xi}|), & \text{when } f(z) \geq 0 \text{ and } \boldsymbol{\xi} \in \boldsymbol{B}_\delta, \\ \leq f(z) + g(|\boldsymbol{\xi}|), & \text{when } f(z) \leq 0 \text{ and } \boldsymbol{\xi} \in \boldsymbol{B}_\delta, \end{cases} \quad (8.1.3)$$

where f *is continuous and non-decreasing in* $\mathbb{R}$, *and* g *is continuous in* $\boldsymbol{B}_\delta$ *with* $g(0) = 0$.

(i) *If* f *has only a single zero,* γ, *then the only entire solution* $u \in C^2(\mathbb{R}^n)$ *of* (8.1.2) *such that* $u(x) = o(|x|)$ *as* $|x| \to \infty$ *is* $u \equiv \text{const.} = \gamma$.

(ii) *If* f *has no zeros, then there are no entire solutions* $u \in C^2(\mathbb{R}^n)$ *of* (8.1.2) *such that* $u(x) = o(|x|)$ *as* $|x| \to \infty$.

Equation (8.1.1) is obviously covered by Theorem 8.1.1.

Proof. Case (i). Let $x_0 \in \mathbb{R}^n$ and $c = u(x_0)$. We assert that $f(c) \leq 0$. Otherwise suppose for contradiction that $f(c) > 0$. For $\varepsilon \in (0, \delta)$ put

$$v(x) = u(x) - c - \varepsilon h(x), \qquad h(x) = \sqrt{1 + |x - x_0|^2} - 1.$$

Then $v(x_0) = 0$, while $v(x) \to -\infty$ as $|x| \to \infty$. Consequently v takes a non-negative maximum at some point y. Hence $v(y) = u(y) - c - \varepsilon h(y) \geq 0$, so $u(y) > c$ and $f(u(y)) \geq f(c) > 0$ since f is non-decreasing. Moreover,

$$Dv(y) = Du(y) - \varepsilon Dh(y) = \mathbf{0}, \qquad a_{ij}(y, u(y), Du(y)) \partial^2_{x_i x_j} v(y) \leq 0.$$

Since $|Du(y)| = \varepsilon |Dh(y)| \leq \varepsilon < \delta$, by evaluating (8.1.2) at y and using (8.1.3) we get

$$
\begin{aligned}
f(c) - g(\varepsilon |Dh(y)|) &< f(u(y)) - g(|Du(y)|) \\
&\leq -B(y, u(y), Du(y)) = a_{ij}(y, u(y), Du(y)) \partial^2_{x_i x_j} u(y) \qquad (8.1.4) \\
&\leq \varepsilon a_{ij}(y, u(y), Du(y)) \partial^2_{x_i x_j} h(y) \leq \varepsilon \sum_{i=1}^{n} a_{ii}(y, u(y), \varepsilon Dh(y)),
\end{aligned}
$$

since $\partial^2_{x_i x_j} h(x) = (1 + |x - x_0|)^{-1/2} \delta_{ij} - (1 + |x - x_0|)^{-3/2} (x_i - x_{0,i})(x_j - x_{0,j})$ and $[a_{ij}]$ is non-negative definite. Thus, letting $\varepsilon \to 0$ in (8.1.4) yields $f(c) \leq 0$, a contradiction. Thus $f(c) \leq 0$.

In the same way we find $f(c) \geq 0$, so $f(c) = 0$ and $c = \gamma$. This completes the proof of (i).

Case (ii). Suppose first that $f(z) > 0$ for all z. Then exactly as in (i) we find that $f(c) \leq 0$ for any value c in the range of u. The existence of an entire solution such that $u(x) = o(|x|)$ as $|x| \to \infty$ therefore leads to a contradiction. The case when $f(z) < 0$ for all z is treated similarly. □

To prove that (I) has no entire solutions which are $o(|x|)$ as $|x| \to \infty$, observe from Theorem 8.1.1, with $B(x, z, \boldsymbol{\xi}) = -f(z)$, that $f(c) = 0$ for all c in the range of u. Thus in fact $\Delta u = 0$. But then (making use of the spherical harmonic expansion of u about a given origin) we see that $u \equiv$ constant, as required.

To prove (II) let $f(z) = (9/16)z$ and $g(|\boldsymbol{\xi}|) = 0$. Then (8.1.3) applies with $\delta = 1/2$. Hence by (i) we find that $u \equiv 0$ is the only entire solution such that $u(x) = o(|x|)$ as $|x| \to \infty$. To obtain (IV), we apply Theorem 8.1.1 (ii) with $f(z) \equiv a$, $a \neq 0$, and $g(|\boldsymbol{\xi}|) = |\boldsymbol{\xi}|^2$. Thus there can be no entire bounded solution, or even an entire solution such that $u(x) = o(|x|)$ as $|x| \to \infty$.

As the examples (I)–(IV) make clear, there seems no simple overall Liouville theorem for quasilinear elliptic equations, even in cases in which the principal part consists of the Laplace operator. Nevertheless, there are further interesting results which can be obtained without difficulty.

A first case of interest occurs if f is strictly monotone in $\mathbb{R}$. Then f has at most one zero in $\mathbb{R}$, and in turn every entire solution which is $o(|x|)$ as $|x| \to \infty$ is constant. An important example is the *capillary surface equation*

$$\text{div}\left(\frac{Du}{\sqrt{1+|Du|^2}}\right) = \kappa\, u, \qquad \kappa > 0. \tag{8.1.5}$$

In particular, the only entire solution which is $o(|x|)$ as $|x| \to \infty$ is $u \equiv 0$. In fact, as will be seen later, the only entire solution of (8.1.5) which has at most algebraic growth at infinity is $u \equiv 0$.

Even more, the result of Theorem 8.1.1 extends to solutions u defined in exterior domains, the result being again that $f(c) = 0$ for all values c which the solution u can attain at ∞, see Problems 8.3 (i) and 8.4. For (8.1.5), this means that any exterior capillary surface solution must approach the limit 0 as $|x| \to \infty$, if it is algebraic as $|x| \to \infty$.

When $f = f(z)$ is non-decreasing but not strictly monotone in z, it is still possible to draw useful conclusions. Suppose for example that f vanishes for all $z \le 0$ and is non-decreasing and positive for $z > 0$. In this case the proof of Theorem 8.1.1 supplies the conclusion that all solutions of (8.1.2) whose positive part is $o(|x|)$ as $|x| \to \infty$ must be non-negative.

Furthermore, by choosing other functions h than that used in the proof of Theorem 8.1.1 we can obtain significant extensions of this result. For example, if $h(x) \to \infty$ as $|x| \to \infty$ and

$$\varepsilon a_{ij}(y, u(y), \varepsilon Du(y))\partial^2_{x_i x_j} h(y) \le \text{const.}\, \varepsilon^\beta, \qquad \beta > 0, \tag{8.1.6}$$

then Theorem 8.1.1 continues to hold provided $g(s) \equiv 0$ and $u(x) = o(h(x))$ as $|x| \to \infty$.

For example, the following result holds for the p-Laplace operator.

Theorem 8.1.2. *Let $u \in C^1(\mathbb{R}^n)$, with also $u \in C^2$ in the neighborhood of any point y where $Du \ne \mathbf{0}$, be an entire (distribution) solution of*

$$\Delta_p u = f(u), \qquad p > 1, \tag{8.1.7}$$

such that $u(x) = o(|x|^{p'})$ as $|x| \to \infty$.

Assume that f is a non-decreasing function which does not vanish identically. Then $u \equiv \text{constant}$.

The case $p = 2$ of this result is due to A. Farina [35].

Proof. Writing (8.1.7) in standard form it becomes

$$a_{ij}(Du)\partial^2_{x_i x_j} u(y) = f(u),$$

where

$$a_{ij}(Du) = |Du|^{p-2}\delta_{ij} + (p-2)|Du|^{p-4}\partial_{x_i} u \partial_{x_j} u.$$

Now take $h(x) = |x|^\alpha$, $\alpha > 1$, so that, after a short calculation,

$$\varepsilon a_{ij}(\varepsilon Dh(y))\partial^2_{x_i x_j} h(y) = (\varepsilon\alpha)^{p-1}\{n + \alpha(p-1) - p\}|y|^{\alpha(p-1)-p}.$$

Taking $\alpha = p/(p-1) = p'$ then gives

$$\varepsilon a_{ij}(\varepsilon Dh(y))\partial^2_{x_i x_j} h(y) = (\varepsilon p')^{p-1} n,$$

so (8.1.6) is valid with $\beta = p - 1$.

Since $f \not\equiv 0$, we may suppose for definiteness that $\gamma = \{\text{greatest zero of } f\} > 0$. Let $u(x_0) = c > \gamma$, with $Du(x_0) \neq \mathbf{0}$. Then following the proof of Theorem 8.1.1 (i), we find $Du(y) = \varepsilon Dh(y) \neq \mathbf{0}$ if $y \neq x_0$; that is, in all cases, $u \in C^2$ in a neighborhood of y. In turn we get $f(c) \leq (\varepsilon p')^{p-1} n$. Letting $\varepsilon \to 0$ and noting that $f(c) > 0$ then gives a contradiction. That is $u(x_0) \leq \gamma$. From this, it follows easily by continuity that $u(x) \leq \gamma$ for all $x \in \mathbb{R}^n$.

Similarly $u \geq \gamma'$, where γ' is the least zero of f (or $\gamma' = -\infty$ if $f \equiv 0$ for $z \leq \gamma$). Finally, since f is non-decreasing we find also $f(z) \equiv 0$ when $z \in (\gamma', \gamma)$. In summary, the solution u is necessarily bounded on one side, with $f(u(x)) = 0$ for all values $u(x)$ in the range of the solution, that is,

$$\Delta_p u = 0.$$

The Liouville theorem, Corollary 7.2.3, now implies $u \equiv$ constant, completing the proof. $\qquad\square$

Similar ideas can be applied to the case of the mean curvature operator, leading to the surprising

Theorem 8.1.3. *Let $u \in C^2(\mathbb{R}^n)$ be an entire solution of*

$$\mathrm{div}\left(\frac{Du}{\sqrt{1+|Du|^2}}\right) = f(u). \tag{8.1.8}$$

such that u has at most algebraic growth as $|x| \to \infty$.

Suppose that f is non-decreasing and does not vanish identically. Then $u \equiv$ constant.

The proof is essentially the same as before, though with two main differences. First, one shows that if $h(x) = |x|^\alpha$, $\alpha > 1$, then with

$$a_{ij}(Du) = (1 + |Du|^2)^{-1/2}\delta_{ij} - (1 + |Du|^2)^{-3/2}\partial_{x_i}u\,\partial_{x_i}u,$$

one gets

$$\varepsilon a_{ij}(y, u(y), \varepsilon Du(y))\partial^2_{x_i x_j}h(y) \le \varepsilon^{1/\alpha} \qquad (8.1.9)$$

provided that ε is suitably small (see Problem 8.4).

Then as in the proof of Theorem 8.1.2 one finds that u is bounded on at least one side, and that

$$\operatorname{div}\left(\frac{Du}{\sqrt{1 + |Du|^2}}\right) = 0.$$

Finally by a result of Bombieri, De Giorgi and Miranda [15] necessarily u is constant.

Notes

The conclusions of this section are in most respects new, though based originally on [96]. Other related results can be found in [70] and in the extensive monograph [35].

It has been assumed throughout the section that f is a non-decreasing function of u. When this is not the case, for example for the equation

$$\Delta u + |u|^{q-2}u = 0, \qquad q > 1,$$

the situation is entirely different and the results much more delicate (moreover, for the most part, being independent of maximum principle techniques). There is a large literature concerning this case, cf. [12], [13], [41], [60], and particularly [74] and [101], to which the reader can be referred.

8.2   Radial symmetry

Let B be a ball in $\mathbb{R}^n$, for definiteness centered at the origin, and consider the Dirichlet problem

$$\Delta u + f(u) = 0, \qquad u > 0 \quad \text{in } B,$$
$$u = 0 \quad \text{on } \partial B. \qquad\qquad (8.2.1)$$

One may expect the existence of radial solutions $u = u(r)$ of this problem, coming from the ordinary differential equation

$$u'' + \frac{n-1}{r} u' + f(u) = 0.$$

The question then arises whether solutions are *necessarily* radial. Delicate examples show that this in fact may not be the case, see for example [38, page 104]. On the other hand, if the function f is, say, of class C^1, then the answer is yes, as a consequence of the following

Theorem 8.2.1 (Radial Symmetry). *Let B be an open ball in $\mathbb{R}^n$, $n \geq 1$. Assume $u \in C^2(B) \cap C(\overline{B})$ satisfies (8.2.1), where f is locally Lipschitz continuous in $\mathbb{R}_0^+$. Then u is radially symmetric, that is can be written in the form $u = u(r)$, $r = |x|$.*

This result is due to Gidas, Ni and Nirenberg [40] for solutions of class $C^2(\overline{B})$ and to Berestycki and Nirenberg [11] for the stated case. A short proof of Theorem 8.2.1 was given by Brezis [16].

Theorem 8.2.1 allows extension to radially symmetric quasilinear equations, moreover without the assumption of positivity of the solution, or the full Lipschitz continuity of the nonlinearity f. There are two main cases, first when the solution $u \in C^1(\overline{B})$, and second for $u \in C^1(B) \cap C(\overline{B})$.

In the second result, which we state as Theorem 8.2.3, less regularity is required of u near the boundary of B. This however leads to stronger regularity hypotheses being needed for the operator A and nonlinearity f. At the same time, it is easy to see that these extra hypotheses automatically hold for the problem (8.2.1), where $A(z, s) \equiv 1$ and $f = f(z)$. Thus Theorem 8.2.1 is a special case of Theorem 8.2.3.

Theorem 8.2.2 (Radial Symmetry, I). *Let B be an open ball in $\mathbb{R}^n$, $n \geq 1$. Assume $u \in C^1(\overline{B})$ is a distribution solution of the problem*

$$\mathrm{div}\{A(u, |Du|)Du\} + f(u, |Du|) = 0, \qquad u \geq 0 \quad in\ B,$$
$$u = 0 \quad on\ \partial B. \tag{8.2.2}$$

Here $A = A(z, s) : \mathbb{R}_0^+ \times \mathbb{R}_0^+ \to \mathbb{R}^+$ is assumed continuously differentiable with

$$sA'(z, s) + A(z, s) > 0 \qquad (\ ' = \partial_s); \tag{8.2.3}$$

while the function $f = f(z, s)$ is locally Lipschitz continuous in $\mathbb{R}_0^+ \times \mathbb{R}_0^+$.

Then u is radially symmetric about the origin in B and is of class $C^2(B)$. When $n \geq 2$, then either $u \equiv 0$ or $u > 0$ in B with $u'(r) < 0$ for $0 < r < R$.

The principal operator in (8.2.2) is closely related to the variational integral

$$I[u] = \int_{\Omega} \mathscr{G}(u, |Du|) \, dx,$$

where $\mathscr{G}$ and A are related by $A(z, s) = \mathscr{G}'(z, s)/s$, $s > 0$. Ellipticity then is equivalent to $\mathscr{G}''(z, s) > 0$ for $s > 0$. Theorem 8.2.2 applies in particular to the mean curvature equation

$$\mathrm{div} \left(\frac{Du}{\sqrt{1 + |Du|^2}} \right) = f(u, |Du|).$$

Here $A = A(s) = (1 + s^2)^{-1/2} > 0$ and $A(s) + sA'(s) = (1 + s^2)^{-3/2} > 0$, that is the equation is elliptic. Thus *every solution in $C^1(\overline{B})$ with boundary condition $u = 0$ on ∂B is radially symmetric.*

Theorem 8.2.3 (Radial Symmetry, II). *Let B be an open ball in $\mathbb{R}^n$, $n \geq 1$. Assume $u \in C^1(B) \cap C(\overline{B})$ is a distribution solution of the problem (8.2.2). Here the operator $A = A(z, s) : \mathbb{R}_0^+ \times \mathbb{R}_0^+ \to \mathbb{R}^+$ is assumed to be uniformly continuously differentiable in $\Gamma \times \mathbb{R}_0^+$, where Γ is any compact subset of $\mathbb{R}_0^+$, with both quantities*

$$A(z, s), \qquad sA'(z, s) + A(z, s) \tag{8.2.4}$$

uniformly bounded away from zero in $\Gamma \times \mathbb{R}_0^+$; while the function $f = f(z, s)$ is uniformly Lipschitz continuous in $\Gamma \times \mathbb{R}_0^+$. Then the conclusion of Theorem 8.2.2 continues to hold.

Condition (8.2.4) can be expressed alternatively as stating that the differential equation is uniformly elliptic.

Remark. When the restriction $u \geq 0$ in B in Theorems 8.2.2 and 8.2.3 is strengthened to $u > 0$ in B, it is not hard to see from the proofs below that $f(z, s)$ does not need to be lower Lipschitz continuous in the variable z at $z = 0$, though upper Lipschitz continuity is still required. This allows for example the interesting class of nonlinearities $f = f(z, s)$ having asymptotic form $-z^q$ near $z = 0$, with $0 < q < 1$, not previously noted in the literature.

The possibility of radial symmetry on annuli is the concern of the next result.

Theorem 8.2.4. *Let B be a ball or an annulus $B = B_2 \setminus \overline{B}_1$, centered at the origin. Assume that $u \in C^1(B) \cap C(\overline{B})$ is a solution of the problem*

$$\mathrm{div}\{\rho(r)A(|Du|)Du\} + f(r, u) = 0 \qquad \text{in } B,$$
$$u = \text{constant} \quad \text{on any component of } \partial B.$$
$$(8.2.5)$$

Here the function A is assumed to be positive and $s \mapsto sA(s)$ strictly increasing in $\mathbb{R}^+$, with $sA(s) \to 0$ as $s \to 0$; while $f = f(r, z)$, $r = |x|$, is locally bounded in $B \times \mathbb{R}$, and non-increasing in z; finally the function ρ is positive and locally bounded in $B \setminus \{0\}$.

$\qquad$ *Then u is unique and radially symmetric.*

$\qquad$ In contrast with Theorem 8.2.2, no restriction on the sign of u is required in Theorem 8.2.4, and even more in Theorem 8.2.4 *the operator A can be singular*, e.g., $A(s) = s^{p-2}$, $p > 1$, whereas in Theorem 8.2.2 necessarily $A(z, 0) > 0$. On the other hand, the monotonicity condition on f, replacing locally Lipschitz continuity in Theorem 8.2.2, is itself a strong requirement.

Proof of Theorems 8.2.2–8.2.4

Proof of Theorem 8.2.2. We use the technique of moving planes, introduced in [2] and [95].

$\qquad$ Write $x = (x_1, x')$ with $x' = (x_2, \ldots, x_n)$. For $\lambda \in (0, R)$, where R is the radius of B, we set

$$B_\lambda = \{x \in B : x_1 > \lambda\} \quad \text{and} \quad \tilde{x} = \tilde{x}^\lambda = (2\lambda - x_1, x');$$

$\tilde{x}$ is the reflection of the point x in the hyperplane T with equation $x_1 = \lambda$. Clearly $\tilde{x} \in B$ when $x \in B_\lambda$, so we can define

$$v = v^\lambda(x) = u(\tilde{x}).$$

It is easy to see that v, along with u, satisfies

$$\mathrm{div}\{A(|Du|)Du\} + f(u, |Du|) = 0 \quad \text{in } B_\lambda.$$

Hence in B_λ

$$\mathrm{div}\{A(v, |Dv|)Dv - A(u, |Du|)Du\} + f(v, |Dv|) - f(u, |Du|) = 0. \quad (8.2.6)$$

Put $w = w^\lambda = v^\lambda - u \in C^1(B_\lambda) \cap C(\overline{B}_\lambda)$. Then $w \geq 0$ on ∂B_λ, that is both on $\partial B_\lambda \cap \{x_1 > \lambda\}$ and on $B \cap \{x_1 = \lambda\}$.

It follows from (8.2.3) that the matrix $[\partial_{\boldsymbol{\xi}}(A(z, |\boldsymbol{\xi}|)\boldsymbol{\xi})]$ is locally positive definite in $\mathbb{R}_0^+ \times \mathbb{R}_0^+$; moreover $\partial_z A$ and $|\partial_{\boldsymbol{\xi}} A|$ are locally bounded in $\mathbb{R}_0^+ \times \mathbb{R}_0^+$. In turn, using the fact that Du is bounded in B, we see that for $x \in B_\lambda$ there holds

$$\begin{aligned} \langle A(v, |Dv|)Dv - A(u, |Du|)Du, \ Dw\rangle &\geq a_1|Dw|^2 - a_2 w^2, \\ |A(v, |Dv|)Dv - A(u, |Du|)Du| &\leq a_3|Dw| + a_4|w|, \end{aligned} \qquad (8.2.7)$$

for appropriate constants a_1, $a_3 > 0$ and a_2, $a_4 \geq 0$: see (2.5.9). Also since $f = f(z, s)$ is locally Lipschitz continuous in $\mathbb{R}_0^+ \times \mathbb{R}_0^+$ we have similarly

$$|f(v, |Dv|) - f(u, |Du|)| \leq b_1|Dw| + b_2 w \qquad (8.2.8)$$

for appropriate constants b_1, $b_2 \geq 0$; the constants in the inequalities (8.2.7) and (8.2.8) obviously depend only on bounds for u and Du in B.

For λ near R, the set B_λ has small measure, e.g., $|B_\lambda| < R - \lambda$. We are therefore in position to apply Theorem 3.3.1. In particular, in view of (8.2.7) and (8.2.8), the equation (8.2.6) takes the form (3.1.1) with w in place of u, and *with* (3.2.1) *holding for $p = 2$*. Since $w \geq 0$ on ∂B_λ it now follows from Theorem 3.3.1 that $w = w^\lambda \geq 0$ in B_λ for λ sufficiently near R. Let

$$\Lambda = \{\lambda \in (0, R) : w^\lambda \geq 0 \text{ in } B_\lambda\}.$$

Thus Λ is non-empty and relatively closed in $(0, R)$.

Let $\lambda \in \Lambda$. Remembering that f is locally *lower* Lipschitz continuous, from the tangency principle Theorem 2.5.2 applied to the pair of solutions u and $v = v^\lambda$ in the set B_λ, we see that either

$$w^\lambda \equiv 0 \quad \text{or} \quad w^\lambda > 0 \quad \text{in } B_\lambda. \qquad (8.2.9)$$

In the sequel we will need the following result.

Lemma 8.2.5. *If $w^\lambda > 0$ in B_λ for all $\lambda \in \Lambda$, then $\Lambda = (0, R)$.*

Proof. Let $\lambda \in \Lambda$. It is obviously enough to show that $\mu \in \Lambda$ when $\mu < \lambda$ and μ is sufficiently near λ. Let K be a compact subset of B_λ with the property that the set $B_\mu \setminus K$ has measure so small that Theorem 3.3.1 applies; this can be accomplished by making at the same time μ suitably near λ. Obviously $w = w^\lambda \geq \delta$ in K for a suitable constant $\delta > 0$. Then for the function w^μ, we have *when $x \in K$*,

$$w^\mu(x) = v^\mu(x) - u(x) = u(\tilde{x}^\mu) - v^\lambda(x) + w^\lambda(x) = u(\tilde{x}^\mu) - u(\tilde{x}^\lambda) + w^\lambda(x) \geq 0,$$

since $|\tilde{x}^\mu - \tilde{x}^\lambda| = 2(\lambda - \mu)$ can be made as small as we wish by taking μ even nearer λ if necessary (and since u is uniformly continuous in B).

In particular, $w^\mu \geq 0$ on ∂K, so in turn

$$w^\mu \geq 0 \quad \text{on } \partial(B_\mu \setminus K) = \partial K \cup \partial B_\mu.$$

Hence by Theorem 3.3.1 we get $w^\mu \geq 0$ in $B_\mu \setminus K$, and in combination $w^\mu \geq 0$ in B_μ. Hence $\mu \in \Lambda$ for all $\mu < \lambda$ which are sufficiently near λ, as required. Thus $\Lambda = (0, R)$. $\qquad\square$

The proof now divides into three cases.

Case 1. $u > 0$ in B. It is easy to see that $w^\lambda > 0$ in B_λ for all $\lambda \in \Lambda$: otherwise, $w_\lambda \equiv 0$ for some $\lambda \in \Lambda$, so in particular we would have $w^\lambda = v^\lambda = 0$ on $\partial B_\lambda \cap \partial B$. But this requires that $u = 0$ on the reflection of ∂B in the hyperplane $x_1 = \lambda$, contradicting the assumption that $u > 0$ in B. It now follows from Lemma 8.2.5 that $\Lambda = (0, R)$ and so $w^\lambda \geq 0$ in B_λ for $0 < \lambda < R$.

By continuity $u(\tilde{x}) - u(x) \geq 0$ for $\lambda = 0$, that is

$$u(x_1, y) \leq u(-x_1, y), \qquad x_1 > 0.$$

The same argument applies with a moving plane $x_1 = \lambda < 0$, with $\lambda \in (-R, 0)$. Thus $u(x_1, y) \leq u(-x_1, y)$, $x_1 < 0$. Consequently $u(x_1, y) = u(-x_1, y)$, and u is symmetric across the hyperplane $x_1 = 0$. By rotation of coordinates the same conclusion applies in all directions and u is symmetric across any hyperplane through the origin. Thus u is radially symmetric.[2]

Case 2. $u \not> 0$ in B, and $n \geq 2$. We assert that there is some $\lambda \in \Lambda$ such that $w^\lambda \equiv 0$ in B_λ. Otherwise, recalling the dichotomy (8.2.9), if $w^\lambda > 0$ for all $\lambda \in \Lambda$, then by Lemma 8.2.5 we would have $\Lambda = (0, R)$. In fact, this is impossible: let $x_0 \in B$ be such that $u(x_0) = 0$, and choose λ so that x_0 lies in the reflection of B_λ across the hyperplane $x_1 = \lambda$. Then at the reflected point $\tilde{x}_0 \in B_\lambda$ there would hold

$$0 < w^\lambda(\tilde{x}_0) = u(x_0) - u(\tilde{x}_0) = -u(\tilde{x}_0) \leq 0,$$

a contradiction.

Let $\lambda_0 \in \Lambda$ be such that $w^{\lambda_0} \equiv 0$ in B_{λ_0}. Then necessarily $u = 0$ on the reflection L of ∂B across the hyperplane $T_0 : x_1 = \lambda_0$. Let y be a point in $\partial B \cap T_0$. We reapply the previous moving planes argument, but now with

---

[2]If one assumes to begin with $u > 0$ in B, as in Theorem 8.2.1, then one can skip the delicate Cases 2 and 3 which follow.

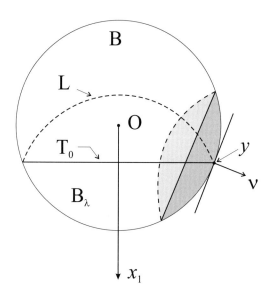

Figure 8.1: The dashed set L is the reflection in the hyperplane T_0 of ∂B. By construction $u = 0$ on L and consequently by the moving plane argument also $u = 0$ in the shaded "*lens*" set Σ.

hyperplanes parallel to the tangent hyperplane to ∂B at y. See Figure 1. The previous reflection and thin set arguments then supply the conclusion that the (new) functions w^λ are identically zero for all λ suitably near R; that is, for these functions the inequality $w^\lambda = w^\lambda(x) = u(x) - u(\tilde{x}) > 0$ is incompatible with the condition $u = 0$ on L. But then $u = 0$ on the boundary of any "*lens*" set Σ for which λ is near R. Hence in turn $u \equiv 0$ at *all* points in any sufficiently small "*lens*" set adjacent to y. Evaluating the main equation (8.2.2) in this set then yields $f(0,0) = 0$.

In turn, (8.2.8) gives

$$f(u, |Du|) \geq -b_1|Du| - b_2 u \qquad \text{for } u \geq 0.$$

The tangency principle Theorem 2.5.2 then implies $u \equiv 0$ in B; that is, u is (trivially) radially symmetric.

Case 3. $u \not\equiv 0$ *in* B *and* $n = 1$. In this case there exists some point in the interval $B = (-R, R)$ where $u = u' = 0$. There are two subcases.

First, if $f(0,0) \geq 0$, then again by the strong maximum principle one gets $u \equiv 0$ in B.

The remaining case $f(0,0) < 0$ is more complicated. Since this condition makes it impossible to have any subintervals of B where $u \equiv 0$, necessarily B must consist of a finite or denumerable set of open intervals I on which $u > 0$, separated by points where $u = u' = 0$. Consider any such subinterval $I = (a, b)$. On I, u must be a solution of the ordinary differential equation

$$\{A(u, |u'|)u'\}' + f(u, |u'|) = 0. \tag{8.2.10}$$

Then by Case 1 it follows that u must be symmetric about the midpoint of I, with $u' \le 0$ to the right of the midpoint. Using Lemma 8.2.6 below, for the case $n = 1$ and with $J = (\frac{1}{2}(a+b), b)$, we get $u' < 0$ in J; even more, the function u, being a solution in J of the end value problem $u(b) = u'(b) = 0$ with $u' < 0$, must equal, up to translation, a unique function $U(x)$, and the interval J must have a unique length, say d. That is, the difference $b-a = 2d$ must be independent of I, and the solutions for different subintervals I must be identical following translation.

It follows that there are only a finite number of subintervals I, that $2R$ must be a multiple of $b - a$, and finally that u is symmetric on B, though of course consisting of more than a single "*hill*".

To complete the proof of Theorem 8.2.2, it thus remains only to show that when $u > 0$ in B, then the solution $u = u(r)$ obeys $u'(r) < 0$ for $0 < r < R$. To accomplish this, we first observe, since $\Lambda = (0, R)$, that necessarily $u = u(r)$ is non-increasing, hence $u'(r) \le 0$. That equality cannot occur is a consequence of the following

Lemma 8.2.6. *Let J denote the interval $(0, S)$ and let $u \in C^1(J) \cap C(\overline{J})$ be a solution of the ordinary differential equation*

$$\{A(u, |u'|)u'\}' + \frac{n-1}{r} A(u, |u'|)u' + f(u, |u'|) = 0, \quad u > 0, \tag{8.2.11}$$

where A and f satisfy the hypotheses of Theorem 8.2.2. Suppose $u' \le 0$ in J and $u(R) = 0$. Then $u \in C^2(J)$ and $u' < 0$.

Moreover, when $n = 1$ there cannot be more than one value S and one solution $u \in C^1(\overline{J})$ such that $u(S) = u'(S) = 0$ and $u'(r) \le 0$ in J.

Proof. Define $\Phi(z, s) = sA(z, s)$ for $z > 0$, $s \ge 0$. By (8.2.3) the function $\Phi(z, \cdot)$ has a continuously differentiable inverse $\Phi^{-1}(z, \cdot)$. Put

$$v = v(r) = \Phi(u(r), |u'(r)|), \quad r \in J. \tag{8.2.12}$$

Then we can rewrite (8.2.11) in the form (where v' is a weak derivative)

$$\begin{cases} u' = -\Phi^{-1}(u, v), \\ v' = -\dfrac{n-1}{r}\Phi(u, \Phi^{-1}(u, v)) + f(u, \Phi^{-1}(u, v)). \end{cases} \quad (8.2.13)$$

Since $v \in C(J)$ it follows from the second equation of (8.2.13) that in fact $v' \in C(J)$, and in turn, from the first equation of (8.2.13), that $u' \in C^1(J)$ and $u \in C^2(J)$.

If at some point $c \in J$ we have $u(c) = u_0$ and $u'(c) = 0$, then also $u''(c) = 0$ since u' has a maximum at c. But then (8.2.12) gives $v'(c) = 0$, and by (8.2.13) also

$$f(u_0, 0) = 0.$$

This being shown, by the uniqueness of the initial value problem for (8.2.13), for the initial point $r = c$, we get $u \equiv u_0$, $v \equiv 0$, a contradiction since $u(R) = 0$ and $u_0 > 0$ by (8.2.11). That is, $u'(r) > 0$ in J.

The final part of the lemma follows from the uniqueness of the initial (end) value problem together with the translation invariance of (8.2.13) for the case $n = 1$. $\qquad \square$

Proof of Theorem 8.2.3. This is almost the same as for Theorem 8.2.2, the only difference being in the derivation of the estimates (8.2.7) and (8.2.8). The uniformity hypotheses however imply that the matrix $[\partial_{\xi}(A(z, |\xi|)\xi)]$ is uniformly positive definite in $\mathbb{R}_0^+ \times \Gamma$. Then with the help of the uniform differentiability of A, the estimates (8.2.7) and (8.2.8) are obtained as before, with the constants in both inequalities depending only on bounds for u in B. $\qquad \square$

The key technical components in the proof of Theorems 8.2.2 and 8.2.3 are the tangency principle Theorem 2.5.2 and the thin set Theorem 3.3.1. The latter result is relatively straightforward and even applies for solutions in $W^{1,2}(\Omega) \cap C(\overline{\Omega})$. Theorem 2.5.2, on the other hand, is based on the Harnack inequality (2.5.3), and consequently is a considerably deeper result. At the same time, (2.5.3) also applies when the solution is in $W^{1,2}(\Omega) \cap C(\overline{\Omega})$, see Theorem 7.1.2.

From these comments, it follows that Theorem 8.2.3 continues to hold for solutions in $W^{1,2}(\Omega) \cap C(\overline{\Omega})$. As a special case, Theorem 8.2.1 remains valid when u is of class $W^{1,2}(\Omega) \cap C(\overline{\Omega})$, as observed by Dancer [26] for the case of the Laplace operator.

More elementary proofs of Theorems 8.2.2 and 8.2.3 can be given if $u \in C^2(\Omega) \cap C(\overline{\Omega})$, for then we can use the tangency principle Theorem 2.2.1, based strictly on the Hopf strong maximum principle.

Proof of Theorem 8.2.4. Consider a second solution $v(x) = u(-x_1, y)$. Since v is equally a solution of (8.2.5), it follows from Theorem 2.6.2 that $u \equiv v$ in B. That is, u must be symmetric across the plane $x_1 = 0$. But then as in the proof of Theorem 8.2.2, the solution must be radial.   □

Notes

For the problem (8.2.1), Fraenkel [38, Theorem 3.6] gives conditions on f closely related to those indicated in the remark after Theorem 8.2.3. Castro and Shivaji [18] removed the positivity condition on the solution u in (8.2.1) in the case $n \geq 2$. Theorem 8.2.4 is Theorem 1.1 of [78].

The (complete) symmetry results of Theorems 8.2.2 and 8.2.4 can easily be extended to unidirectional symmetry for domains which exhibit symmetry in only one (or several) directions, the proofs being essentially unchanged from the radial case. A summary of results of this type is given in [16]; see also [11], [38], [66], [75].

Other work of interest, e.g., for radial symmetry when $\Omega = \mathbb{R}^n$, or for degenerate operators, is contained in the papers [24], [25], [33], [100].

The reader can also be referred to the Notes for Chapter 3 of [38].

8.3 Symmetry for overdetermined boundary value problems

In this section we consider overdetermined boundary value problems on a general domain Ω, for example when the boundary conditions involve both Dirichlet and Neuman data. In this case, a natural question is whether the domain itself must be restricted.

To be specific, let Ω be a bounded domain of $\mathbb{R}^n$, $n \geq 2$, having a smooth boundary $\partial\Omega$. Suppose as a first prime example the Poisson differential equation

$$\Delta u + 1 = 0 \qquad \text{in } \Omega, \tag{8.3.1}$$

together with the overdetermined boundary conditions

$$u = 0, \qquad \partial_\nu u = \text{const.} \quad \text{on } \partial\Omega. \tag{8.3.2}$$

Must Ω be a ball? We shall show that the answer is affirmative, and that u must have the specific form $(R^2 - r^2)/2n$, where R is the radius of the ball and r denotes distance from its center. The precise result is as follows.

Theorem 8.3.1. *Let Ω be a bounded domain with boundary of class C^2. Suppose there exists a solution $u \in C^2(\overline{\Omega})$ of the overdetermined problem (8.3.1)–(8.3.2). Then Ω is a ball and u has the specific form $(R^2 - r^2)/2n$ noted above.*

For the physical motivation of Theorem 8.3.1, consider a viscous incompressible fluid moving in straight parallel streamlines through a straight pipe of given cross sectional form Ω. If we fix rectangular coordinates in space with the z-axis directed along the pipe, it is well known that the flow velocity u along the pipe is then a function of x, y alone, satisfying the Poisson differential equation

$$\Delta u + \kappa = 0 \qquad \text{in } \Omega \subset \mathbb{R}^2,$$

where κ is a constant related to the viscosity and density of the fluid and to the pressure differential per unit length along the pipe. Supplementary to the differential equation one has the adherence condition $u = 0$ on $\partial\Omega$. Finally, the tangential stress per unit area on the pipe wall is given by the quantity $\mu\partial_\nu u$, where μ is the viscosity. Theorem 8.3.1 states that *the tangential stress on the pipe wall is the same at all points of the wall if and only if the pipe has a circular cross section.*

Exactly the same differential equation and boundary conditions arise in linear theory of torsion of a solid straight bar of cross section Ω. Theorem 8.3.1 then states that, *when a solid straight bar is subject to torsion, the magnitude of the resulting traction which occurs at the surface of the bar is independent of position if and only if the bar has a circular cross section.*

Theorem 8.3.1 is a special case of the following general result for quasilinear equations.

Theorem 8.3.2. *Suppose the functions $A = A(z,s)$ and $f = f(z,s)$ satisfy the hypotheses of Theorem 8.2.2, but with A now being assumed twice continuously differentiable in $\mathbb{R}_0^+ \times \mathbb{R}_0^+$.*

Figure 8.2: Liquid rise in a non-circular capillarytube. Here γ is the wetting angle.

Let $u \in C^2(\overline{\Omega})$ be a solution of the problem

$$\operatorname{div}\{A(u, |Du|)Du\} + f(u, |Du|) = 0, \qquad u > 0 \quad \text{in } \Omega,$$
$$u = 0, \qquad \partial_\nu u = \text{constant} \quad \text{on } \partial\Omega, \tag{8.3.3}$$

where Ω is a bounded domain with boundary of class C^2. Then Ω is a ball, and u is radially symmetric about its center.

The proof is given below. With the help of Theorem 8.3.2, we can consider the case of a liquid rising in a straight capillary tube of cross section $\Omega \subset \mathbb{R}^2$. The function $u = u(x, y)$ describing the upper surface of the liquid satisfies the equation

$$\operatorname{div}\left(\frac{Du}{\sqrt{1 + |Du|^2}}\right) = \kappa\, u,$$

where κ is a positive constant, see Example 2 of the Introduction. The requirement that the wetting angle γ at the wall of the tube be constant leads to the boundary condition

$$\partial_\nu u = \cot\gamma = \text{constant} \quad \text{on } \partial\Omega,$$

where ν is the outward normal direction. Then, *provided the wetting angle γ is different from $\pi/2$, a liquid will rise to the same height at each point of the wall of a capillary tube if and only if the tube has a circular cross section.* See Figure 8.2. When $\gamma = \pi/2$ the unique solution is $u \equiv 0$ for any cross sectional form of the tube.

Remark. The domain Ω in Theorem 8.3.2 need not be assumed simply connected. The conclusion that the domain must be a ball (simply connected) is unaffected.

Proof of Theorem 8.3.2. The idea is the same as for Theorem 8.2.2, using the method of moving planes, but without originally knowing the location of the eventual center of Ω.

Let $\lambda \in \mathbb{R}$ and define as before $\tilde{x} = (2\lambda - x_1, y)$. In general $\tilde{x} \notin \Omega$. Let $\lambda_0 \in \mathbb{R}$ be such that the hyperplane $x_1 = \lambda_0$ is one-sidedly tangent to Ω, that is, with $\Omega \subset \{x \in \mathbb{R}^n : x_1 < \lambda_0\}$. Consider the set

$$\Omega_\lambda = \{x \in \Omega : \lambda < x_1 < \lambda_0\}.$$

Since $\partial\Omega$ is of class C^2, it is evident that at least when λ is suitably near λ_0 and $x \in \Omega_\lambda$, then $\tilde{x} \in \Omega$. Consequently for such λ and for $x \in \overline{\Omega}_\lambda$ we can define $v(x) = v^\lambda(x) = u(\tilde{x})$ and $w = w^\lambda = v^\lambda - u$. Moreover, $w \geq 0$ on $\partial\Omega_\lambda$ as before, and again as before if λ is even closer to λ_0, if necessary, the thin set Theorem 3.3.1 gives $w \geq 0$ in Ω_λ.

Now define

$$\Lambda = \{\lambda < \lambda_0 : x \in \Omega_\lambda \text{ implies } \tilde{x} \in \Omega \text{ and } w^\lambda(x) \geq 0\};$$

of course Λ is non-empty and closed. Consider the set $Q_\lambda = \partial\Omega \cap T_\lambda$, where T_λ is the hyperplane $x_1 = \lambda$, and let ν denote the exterior normal vector to Ω at points of Q_λ. It is evident that $\langle \nu, e_1 \rangle < 0$ when λ is near λ_0, and that as λ decreases there would be a first value λ_1 where $\langle \nu, e_1 \rangle = 0$ for some point $y \in Q_{\lambda_1}$.

Step 1. *Assume* $\lambda_1 < \lambda < \lambda_0$ *and* $\lambda \in \Lambda$. As in Case 1 of the proof of Theorem 8.2.2 we must have $w^\lambda > 0$ in B_λ.

We now consider two subcases:

(i) there is $y \in \partial\Omega \setminus T_\lambda$ such that $\tilde{y} \in \partial\Omega$;
(ii) $\tilde{y} \in \Omega$ for all $y \in \partial\Omega \setminus T_\lambda$.

For case (i) we use the overdetermined condition $\partial_\nu u = c = \text{constant}$ on $\partial\Omega$. In fact $\partial_\nu u(y) = c$, $\partial_\nu v(y) = \partial_\nu u(\tilde{y}) = c$, so that $\partial_\nu w(y) = 0$. Recalling that $w \geq 0$ in Ω_λ, the boundary point Theorem 2.7.1 applied at the boundary point y shows that $w \equiv 0$ in Ω_λ. In turn, $u = 0$ on the reflection of $\partial\Omega_\lambda \cap \partial\Omega$. Since $\langle \boldsymbol{\nu}, \boldsymbol{e}_1 \rangle < 0$ on $\partial\Omega \cap T_\lambda$ it follows that $u = 0$ at a set of interior points of Ω, a contradiction. That is, case (i) cannot occur.

In case (ii), it is apparent by simple geometry that $\tilde{x} \in \Omega$ also for all $x \in \Omega_\mu$, when the value $\mu < \lambda$ is sufficiently near λ. But then we can apply Lemma 8.2.5 of Theorem 8.2.2 to show that Λ is open. That is, $\Lambda = (\lambda_1, \lambda_0)$, and in particular $w^\lambda \geq 0$ in B_λ when $\lambda = \lambda_1$.

Step 2. Let $\lambda = \lambda_1$ and choose $y \in Q_\lambda$ so that $\langle \boldsymbol{\nu}, \boldsymbol{e}_1 \rangle = 0$ at y. Since $w = 0$ on T_λ we have $\partial_\nu w(y) = 0$; because $\partial_t w(y) = 0$ for any direction $\boldsymbol{t}$ tangent to $\partial\Omega$ at y, it follows that $Dw(y) = \boldsymbol{0}$.

We now wish to apply the edge theorem stated as Lemma 2 in [95]. To this end, it is first necessary to write the difference equation (8.2.6) in non-divergence form. In fact, since the function A is twice continuously differentiable in $\mathbb{R}_0^+ \times \mathbb{R}_0^+$, we can write (8.3.3) in the form (after division by A)

$$\tilde{a}_{ij}(u, Du)\partial^2_{x_i x_j} u + \tilde{b}(u, |Du|)|Du|^2 + \tilde{f}(u, |Du|) = 0,$$

where

$$\tilde{a}_{ij}(z, \boldsymbol{\xi}) = \delta_{ij} + h(z, |\boldsymbol{\xi}|)\frac{\xi_i \xi_j}{|\boldsymbol{\xi}|},$$

and

$$h(z, s) = \partial_s A(z, s)/A(z, s),$$
$$\tilde{b}(z, s) = \partial_z A(z, s)/A(z, s),$$
$$\tilde{f}(z, s) = f(z, s)/A(z, s).$$

A similar non-divergence equation of course holds also for v.

Then by subtraction we get, using the Lipschitz continuity of h, $\tilde{b}$ and $\tilde{f}$ in the variables z and s,

$$\tilde{a}_{ij}(v, Dv)\partial^2_{x_i x_j} w \leq b_1 |Dw| + c_1 w$$

and equally (!)

$$\tilde{a}_{ij}(u, Du)\partial^2_{x_i x_j} w \leq b_2 |Dw| + c_2 w.$$

Finally, adding the last two inequalities yields

$$a_{ij}(x)\partial^2_{x_i x_j} w \le b|Dw| + cw, \qquad (8.3.4)$$

where

$$a_{ij}(x) = 2\delta_{ij} + \left\{ h(u, |Du|) \frac{\partial_{x_i} u \, \partial_{x_j} u}{|Du|} + h(v, |Dv|) \frac{\partial_{x_i} v \, \partial_{x_j} v}{|Dv|} \right\}. \qquad (8.3.5)$$

The matrix $[a_{ij}(x)]$ is bounded and strictly elliptic in Ω_λ. Moreover, it has the crucial property

$$a_{1j} = 0 \quad \text{on } T_\lambda \cap \Omega, \qquad j = 2, \dots, n. \qquad (8.3.6)$$

Indeed on T_λ we have, by the reflection construction, $\partial_{x_1} v = -\partial_{x_1} u$, $\partial_{x_j} v = \partial_{x_j} u$ for $j = 2, \dots, n$, $|Dv| = |Du|$, whence (8.3.6) follows from (8.3.5). But also the coefficients a_{ij} are uniformly Lipschitz continuous in Ω_λ, so that (8.3.6) implies

$$|a_{1j}(x)| \le \text{Const.}\, x_1 \quad \text{in } \Omega_\lambda; \qquad (8.3.7)$$

here it is convenient to choose new coordinates so that T_λ is the hyperplane $x_1 = 0$, with $x_1 > 0$ in Ω_λ, while the x_n-axis is in the normal direction $-\nu$ at y.

Since $w^\lambda \ge 0$, we are now in position to apply Lemma 2 of [95] to the inequality (8.3.4), with the single exception that the right side is no longer zero but instead has the form $b|Dw| + c\,w$, a case not directly covered by the lemma. In order not to obstruct the flow of the proof we defer discussion of this point until the Appendix at the end of the section. Recalling that $w(y) = 0$, $Dw(y) = \mathbf{0}$, the conclusion of the lemma is that either $w \equiv 0$ in Ω_λ or $\partial^2_{s^2} w(y) > 0$ along any direction s which enters Ω_λ at y.

In fact, $D^2 w(y) = \mathbb{O}$. To see this, observe that (continuing to use the special coordinates noted above)

$$w = w^\lambda = u(-x_1, x') - u(x_1, x') \quad \text{in } \overline{\Omega}_\lambda.$$

Consequently on T_λ we have

$$\partial^2_{x_1 x_1} w = \partial^2_{x_i x_j} w = 0, \qquad i, j = 2, \dots, n.$$

Moreover, by the boundary condition $u = 0$, $\partial_\nu u = \text{constant}$ on $\partial\Omega$ there holds

$$\partial^2_{x_i x_n} u = 0, \qquad i = 1, \dots, n-1.$$

and the assertion follows. Lemma 2 of [95] therefore shows that $w \equiv 0$ in Ω_λ.

Hence for $x \in \Omega_\lambda$ there holds $u(\tilde{x}) = u(x)$. In particular, by continuity

$$u(\tilde{x}) = u(x) = 0 \qquad \text{for } x \in \partial\Omega_\lambda \setminus T_\lambda.$$

Consequently $\tilde{x} \in \partial\Omega$ since $u > 0$ in Ω. Otherwise stated, the boundary of Ω is symmetric across the hyperplane T_λ, and in turn Ω is symmetric across T_λ. Since, by rotation, this is also true for corresponding hyperplanes T_λ with arbitrary normal directions, it follows that Ω must be convex. But the only convex domains which have the symmetry just noted are balls. □

The condition that $u > 0$ in Ω can be weakened to $u \geq 0$ provided that either the Neumann constant in (8.3.3) is positive (of course it is necessarily non-negative) or $f(0,0) \geq 0$. The details can be left to the reader.

Appendix to Section 8.3

The calculations involved in the proof of Lemma 2 of [95] (see lines 7–24 on page 314 of [95]) are more complicated than one might wish, but still are within reach of pencil and paper.[3] At the same time, there are three further points which need to be made.

(1) The inequality (8.3.7) takes the place of (26) of Lemma 2 of [95]; it is used on line 13 on page 314.

(2) The terms $b|Dw| + cw$ on the right side of (8.3.4) cause no essential new difficulties, once it is observed that, in the notation of [95],

$$z(x) = \left(e^{-\alpha(x_1 - r_1)^2} - e^{-\alpha r_1^2}\right)\left(e^{-\alpha r^2} - e^{-\alpha r_1^2}\right)$$
$$\leq 2\alpha(r_1 - x_1)x_1 e^{-\alpha r_1^2}\left(e^{-\alpha r^2} - e^{-\alpha r_1^2}\right)$$

(by the mean value theorem as on line 17 of page 324). In turn

$$cz(x) \leq 2\alpha c x_1 r_1 e^{-\alpha(r^2 + r_1^2)} \leq 2\alpha c x_1 r_1 e^{-\alpha[r^2 + (x_1 - r_1)^2]}.$$

Therefore in lines 20, 21 the estimate for Lz need be changed only to include the additional term $-2c/\alpha$ in the first set of braces, which leaves the proof essentially unchanged.

---

[3] Fraenkel [38, page 305] remarks that results of the type of Lemma 2 unavoidably involve "*greater complexity*" than standard boundary point theorems. The proof in [95], as extended by the discussion below, should be judged in the context of Fraenkel's remark.

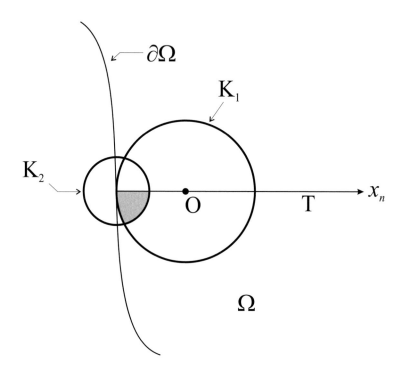

Figure 8.3: *Proof of Lemma 2 of* [95]. The critical point y on $\partial\Omega$, where $\langle \boldsymbol{\nu}, \boldsymbol{e}_1 \rangle = 0$, is at the center of the small ball K_2. The hyperplane T has the equation $x_1 = 0$, with x_1 pointing downward; and the x_n-axis is taken in the direction $-\boldsymbol{\nu}$. (The diagram thus shows an (x_1, x_n)-plane section of $\mathbb{R}^n$ near y.) The ball K_1 has center O on the x_n-axis, is tangent to $\partial\Omega$ at y and $\overline{K_1} \subset \Omega \cup \{y\}$. The radius of K_1 is r_1 and the radius of K_2 is θr_1, with $0 < \theta < 1/2$. The shaded region is the (open) set $K = K_1 \cap K_2 \cap \{x \in \mathbb{R}^n : x_1 > 0\}$.

(3) For lines 22–24 we observe that, again in the notation of [95], see Figure 8.3;

$$z = w = 0 \quad \text{on } T; \qquad z = 0, \qquad w > 0 \quad \text{on } \partial K_1 \cap \partial K',$$
$$z(x) \le 2\alpha r_1 \cdot x_1 \quad \text{on } \partial K_2 \cap \partial K'.$$

By the tangency Theorem 2.5.2 either $w \equiv 0$ or $w > 0$ in Ω_λ. In the latter case, from the boundary point Theorem 2.7.1 with $B(x, z, \boldsymbol{\xi}) = -b|\boldsymbol{\xi}| - cz$, one gets $\partial_{x_1} w > 0$ on $T \cap \Omega$. But because w is continuously differentiable and $w > 0$ in Ω_λ, by compactness it follows that $w \ge \varepsilon x_1$ on $\partial K_2 \cap \partial K'$.

We can now compare the solutions w and mz in K', for suitably small $m > 0$. By the comparison Theorem 2.3.1, noting that the required monotonicity for the function $B(x, z, \boldsymbol{\xi})$ is satisfied since $c \geq 0$, it follows from the fact that $w \geq mw$ on $\partial K'$ (m suitably small) that $w \geq mw$ in K'. Because $Dw = Dz = \mathbf{0}$ at y we obtain, as stated, that $\partial^2_{\boldsymbol{s}^2} w(y) > 0$ along any direction $\boldsymbol{s}$ which enters Ω_λ. $\qquad\square$

Notes

Theorem 8.3.2 is essentially Theorem 2 of [95], the conditions on the non-linearity f however being weaker, and the proof improved over the original version.

The overdetermined boundary value problem for exterior domains when the principal operator is the Laplacian was studied by Reichel [87] and by Aftalion and Busca [1].

8.4 The phenomenon of dead cores

An elliptic equation is said to have a *dead core solution* u in some domain $\Omega \subset \mathbb{R}^n$ provided that there exists an open subset Ω_1 with compact closure in Ω, called the *dead core* of u, such that

$$u \equiv 0 \quad \text{in } \Omega_1, \qquad u > 0 \quad \text{in } \Omega \setminus \overline{\Omega_1}.$$

The condition $u > 0$ could be replaced by $u \neq 0$, but for definiteness (and physical reality) we prefer the condition as stated.

In chemical models, for example, when the values of a solution represent the density of a reactant, the vanishing of a solution then delineates a region (dead core) where no reactant is present (see [5], [29], [81], [82]). We turn to an extended discussion of this phenomenon.

In particular, consider the dead core problem for the model A-Laplace equation

$$\operatorname{div}\{A(|Du|)Du\} - f(u) = 0 \qquad \text{in } \Omega. \tag{8.4.1}$$

The following conditions will be imposed, as in Chapter 1:

(A1) $A \in C(\mathbb{R}^+)$;

(A2) $s \mapsto sA(s)$ *is strictly increasing in* $\mathbb{R}^+$ *and* $\Phi(s) = sA(s) \to 0$ *as* $s \to 0$;

(F1) $f \in C(\mathbb{R})$,

(F2) $f(0) = 0$ *and* f *is non-decreasing in* $\mathbb{R}$.

By the strong maximum principle, Theorem 1.1.1, the equation (8.4.1) can have a dead core only if (1.1.5) fails, that is if $f > 0$ for $u > 0$ and

$$\int_{0+} \frac{ds}{H^{-1}(F(s))} < \infty, \qquad F(u) = \int_0^u F(s)ds, \qquad (8.4.2)$$

with H given by (1.1.4). Consequently, we assume that (8.4.2) holds throughout the sequel, except for Theorems 8.4.2 and 8.4.3.

The equation $\Delta u = |u|^{q-1}u$ for example allows dead cores only if $0 < q < 1$. Actually condition (8.4.2) is not only necessary, but also sufficient for the existence of solutions with dead cores. We have the following main result.

Theorem 8.4.1. *Suppose* $\Phi(\infty) = H(\infty) = \infty$. *Assume the dead core condition* (8.4.2) *holds and let* u *be a* C^1 *distribution solution of* (8.4.1), *with* $0 \leq u(x) \leq m$ *on* $\partial\Omega$ *for some constant* $m > 0$. *Then the following properties are valid:*

(a) $0 \leq u < m$ *in* Ω.

(b) *Assume that*

$$R_0 = \int_0^\infty \frac{ds}{H^{-1}(F(s)/n)} < \infty, \qquad (8.4.3)$$

and let B_R *be a ball with radius* $R \geq R_0$, *compactly contained in* Ω. *Then* u *has a dead core in* Ω *for all* $m > 0$.

(c) *If* Ω' *is any compactly contained set in* Ω, *then* $u \equiv 0$ *in* Ω' *provided that* $m > 0$ *is sufficiently small.*

A more refined version of Theorem 8.4.1 can be obtained when $\Omega = B_R$, where B_R is any open ball in $\mathbb{R}^n$, $n \geq 2$, of radius $R > 0$. Until explicitly noted later, we continue to assume that $\Phi(\infty) = H(\infty) = \infty$.

Theorem 8.4.2. *Let (8.4.2) hold, with $f(z) > 0$ for $z > 0$. Then the problem*

$$\begin{cases} \operatorname{div}\{A(|Du|)Du\} = f(u) & \text{in } B_R, \\ u = m > 0 & \text{on } \partial B_R, \end{cases} \tag{8.4.4}$$

admits a unique C^1 distribution solution u, necessarily radial. Moreover $u = u(r) = u(r,m)$ is of class $C^1[0,R]$ and satisfies $u \geq 0$, $u' \geq 0$ in $[0,R]$ and $u'(0) = 0$, where $' = d/dr$.

Finally, at any $r > 0$ where $u(r,m) > 0$ we have also $u'(r,m) > 0$.

It is easy to see that the solution $u = u(\cdot,m)$ must be of one of the following three types, see Figure 4:

(a) $u > 0$ in B_R;
(b) $u(0,m) = 0$ and $u'(r,m) > 0$ when $r > 0$;
(c) There exists a radius $S \in (0,R)$ such that $u \equiv 0$ in B_S and $u'(r,m) > 0$ in the annulus $S < r < R$.

That is, in case (c) the solution u of (8.4.4) has a dead core B_S. The solution u has further properties of interest.

Theorem 8.4.3. *The function $m \mapsto u(r,m)$ is continuous and non-decreasing in the variable $m\,(>0)$, and $u < m$ in B_R.*

The following theorem gives an important relation between the value m and the existence of dead core solutions of (8.4.4).

Theorem 8.4.4. *Let $u(\cdot,m)$ be the unique solution of (8.4.4). Then either $u(\cdot,m)$ has a dead core for all $m > 0$, or there is a unique (finite) number*

$$m = m_0 = m_0(R) > 0$$

for which a solution $u_0 = u_0(r) = u_0(r,m_0)$ of (8.4.4) in B_R exists, with the properties that

(i) $u_0(0) = 0$;
(ii) $u(0,m) > 0$ for every $m > m_0$;
(iii) $u(\cdot,m)$ has a dead core for every $m \in (0,m_0)$.

For convenience we define $m_0 = m_0(R)$ to be ∞ when $u(0,m) = 0$ for all $m > 0$. The examples

$$\Delta u = (\operatorname{sign} u)\sqrt{|u|}, \tag{8.4.5}$$

$$\Delta_4 u = u \tag{8.4.6}$$

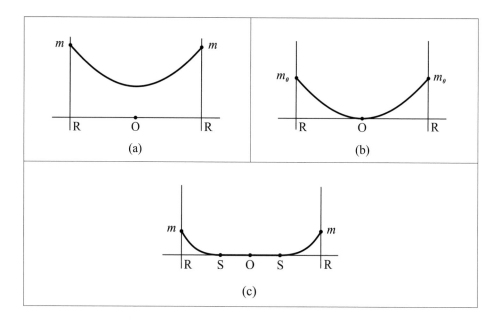

Figure 8.4: Three cases of Theorem 8.4.2. The values m are decreasing from case (a) to case (c).

are particularly interesting as illustrations of the main theorems above. Indeed, both of these are included in the canonical case

$$\Delta_p u = |u|^{q-1} u, \qquad p > 1, \qquad q > 0, \tag{8.4.7}$$

for which $F(u) = |u|^{q+1}/(q+1)$. Here the dead core condition (8.4.2) reduces exactly to

$$0 < q < p - 1.$$

For these special cases, we search for u_0 in the form $c r^k$, c, $k > 0$. Then from (8.4.7) one finds

$$k = \frac{p}{p-1-q}, \qquad c = k^{-k/p'} (n+kq)^{-k/p}, \qquad m_0 = c R^k. \tag{8.4.8}$$

For the case (8.4.5) we have $p = 2$, $q = 1/2$, $k = 4$, so that

$$m_0 = \frac{1}{(n+2)^2} \left(\frac{R}{2} \right)^4,$$

while for (8.4.6) we have $p = 4$, $q = 1$, $k = 2$ and so

$$m_0 = \frac{R^2}{2\sqrt{2(n+2)}}.$$

These reduce exactly to $m_0 = R^4/400$ and $m_0 = R^2/2\sqrt{10}$ when $n = 3$. In particular for the unit radius $R = 1$ we obtain respectively the unexpectedly small numbers $m_0 = 0.00125$ and $m_0 \cong 0.158$.

The equation (8.4.6), when written in full for $n = 2$ has the form

$$|Du|^2 \Delta u + 2(\partial_x u)^2 \partial_{x^2}^2 u + 4\partial_x u \, \partial_y u \, \partial_{xy}^2 u + 2(\partial_y u)^2 \partial_{x^2}^2 u = u,$$

which is analytic in all its variables. Thus dead core behavior is not due simply to a lack of smoothness in the basic equation. In fact (8.4.6) is an analytic partial differential equation, elliptic except when $Du = 0$, which has a non-analytic solution.

As a final example, consider the equation

$$\Delta u = (\text{sign } u)\sqrt{|u|} + |u|^2 u.$$

Here $F(u) = \frac{2}{3}|u|^{3/2} + \frac{1}{4}|u|^4$ so

$$R_0 = \sqrt{\frac{n}{2}} \int_0^\infty \frac{ds}{\sqrt{(2/3)s^{3/2} + s^4/4}} < \infty.$$

By numerical calculation $R_0 \cong 6.4334$ if $n = 2$. Therefore by the results of this section we have $m_0 = \infty$ whenever $R \geq 7$. In particular for the problem

$$\begin{cases} \Delta u = (\text{sign } u)\sqrt{|u|} + |u|^2 u & \text{in } B_7 \subset \mathbb{R}^2, \\ u = m > 0 & \text{on } \partial B_7, \end{cases}$$

a dead core occurs for all $m > 0$. [This result also follows without recourse to numerical calculation, since one can write, when $n = 2$,

$$R_0 = \left(\frac{9}{2}\right)^{1/5} \int_0^\infty \frac{dt}{\sqrt{t^{3/2} + t^4}} < \left(\frac{9}{2}\right)^{1/5} \left\{ \int_0^1 \frac{dt}{\sqrt{t^{3/2}}} + \int_1^\infty \frac{dt}{\sqrt{t^4}} \right\}$$

$$= 5(4.5)^{1/5} \cong 6.75.]$$

The case $n = 3$ can be treated in the same way, with $R_0 \cong 7.879$, so the radius $R = 7$ should be replaced by $R = 8$.

Proof of Theorems 8.4.2 and 8.4.3

Proof of Theorem 8.4.2. Existence of a radial solution u of (8.4.4), with $u \geq 0$, $u' \geq 0$ and $u'(0) = 0$. For the purpose of this proof only, we shall redefine f so that $f(v) = f(m)$ for all $v \geq m$, and $f(v) = 0$ when $v \leq 0$. This will not affect the conclusion of the theorem, since clearly any ultimate solution u of (8.4.4), with $u \geq 0$, $u' \geq 0$ in $[0, R]$, satisfies $0 \leq u \leq m$.

We shall make use of the Leray–Schauder fixed point theorem. Denote by X the Banach space $X = C[0, R]$, endowed with the usual norm $\| \cdot \|_\infty$, and let $\mathcal{T}$ be the mapping from X to X defined pointwise for all $w \in X$ and $r \in [0, R]$ by

$$\mathcal{T}[w](r) = m - \int_r^R \Phi^{-1}\left(s^{1-n} \int_0^s t^{n-1} f(w(t)) dt\right) ds. \tag{8.4.9}$$

Clearly $\mathcal{T}[w](R) = m$. Also

$$\mathcal{T}[w]'(r) = \Phi^{-1}\left(r^{1-n} \int_0^r t^{n-1} f(w(t)) dt\right), \qquad r \in (0, R]. \tag{8.4.10}$$

Obviously $\mathcal{T}[w]'$ is continuous and non-negative in $(0, R]$, since $0 \leq f(w) \leq f(m)$ for all $w \in X$. Moreover $r^{1-n} \int_0^r t^{n-1} f(w(t)) dt$ tends to zero as $r \to 0^+$. Therefore $\mathcal{T}[w]'(r)$ approaches 0 as $r \to 0^+$, since $\Phi(0) = 0$, and in turn $\mathcal{T}[w] \in C^1[0, R]$ with $\mathcal{T}[w]'(0) = 0$.

We claim that if w is a fixed point of $\mathcal{T}$ in X, then $w(0) \geq 0$. Otherwise $w(0) < 0$ and $w(R) = m > 0$. Thus there exists a first point $r_0 \in (0, R)$ such that $w(r) < 0$ in $[0, r_0)$ and $w(r_0) = 0$. Consequently $f(w(r)) = 0$ in $[0, r_0]$ and so $w' \equiv 0$ for $r \in [0, r_0]$ by (8.4.10). Hence $w(r_0) = w(0) < 0$ which is impossible, proving the claim.

Define the homotopy $\mathcal{H} : X \times [0, 1] \to X$ by

$$\mathcal{H}[w, \sigma](r) = \sigma m - \int_r^R \Phi^{-1}\left(\sigma s^{1-n} \int_0^s t^{n-1} f(w(t)) dt\right) ds. \tag{8.4.11}$$

By the above argument, any fixed point $w_\sigma = \mathcal{H}[w_\sigma, \sigma]$ is of class $C^1[0, R]$ and has the properties $w_\sigma \geq 0$, $w_\sigma' \geq 0$ in $[0, R]$ and $w_\sigma(R) = \sigma m$. Additionally, by (8.4.10) we find that $\Phi(w_\sigma') \in C^1[0, R]$, and then from (8.4.9) that w_σ is a classical distribution solution of the problem

$$\begin{cases} [r^{n-1}\Phi(w_\sigma'(r))]' - \sigma r^{n-1} f(w_\sigma(r)) = 0 & \text{in } (0, R], \\ w_\sigma'(0) = 0, \quad w_\sigma(R) = \sigma m. \end{cases} \tag{8.4.12}$$

In turn, it is evident that any function w_1 which is a fixed point of $\mathcal{H}[w, 1]$ (that is $w_1 = \mathcal{H}[w_1, 1]$) is a non-negative radial distribution solution of problem (8.4.4) in $B_R \setminus \{0\}$, with $w'(0) = 0$ and $w' \geq 0$ in $[0, R]$.

Since $f > 0$ for $u > 0$ it follows equally from (8.4.12) that the final statement of the theorem is valid.

We assert that such a fixed point $w = w_1$ exists, using Browder's version of the Leray–Schauder theorem for this purpose (see Theorem 11.6 of [43]).

To begin with, obviously $\mathcal{H}[w, 0] \equiv 0$ for all $w \in X$, that is $\mathcal{H}[w, 0]$ maps X into the single point $w_0 = 0$ in X. (This is the first hypothesis required in the application of the Leray–Schauder theorem.) We show next that $\mathcal{H}$ is compact from $X \times [0, 1]$ into X. First, $\mathcal{H}$ is continuous on $X \times [0, 1]$. Indeed, let $w_j \to w$, $\sigma_j \to \sigma$, $(w_j, \sigma_j) \in X \times [0, 1]$. Then in (8.4.11) clearly $\sigma_j f(w_j) \to \sigma f(w)$, since the modified function f is continuous on $\mathbb{R}$. Hence $\mathcal{H}[w_j, \sigma_j] \to \mathcal{H}[w, \sigma]$, as required.

Next let $(w_k, \sigma_k)_k$ be a bounded sequence in $X \times [0, 1]$. It is clear from (8.4.10) that

$$\|\mathcal{H}[w_k, \sigma_k]'\|_\infty \leq \Phi^{-1}\left(Rf(m)/n\right). \tag{8.4.13}$$

As an immediate consequence of the Ascoli–Arzelà theorem, $\mathcal{H}$ then maps bounded sequences into relatively compact sequences in X, so $\mathcal{H}$ is compact.

To apply the Leray–Schauder theorem it is now enough to show that there is a constant $M > 0$ such that

$$\|w\|_\infty \leq M \quad \text{for all } (w, \sigma) \in X \times [0, 1], \text{ with } \mathcal{H}[w, \sigma] = w. \tag{8.4.14}$$

Let (w, σ) be a pair of type (8.4.14). But, as observed above, one has $w \geq 0$, $w' \geq 0$, so that $\|w\|_\infty = w(R) \leq \sigma m \leq m$. Thus we can take $M = m$ in (8.4.14).

The Leray–Schauder theorem therefore implies that the mapping $\mathcal{T}[w] = \mathcal{H}[w, 1]$ has a fixed point $w \in X$, which is the required solution of (8.4.4) in $B_R \setminus \{0\}$, proving the assertion above.

The fixed point $u = w$ is a C^1 distribution solution of (8.4.4) in B_R. The proof is standard. Let $\varphi \in C_c^1(B_R)$. We have to show that

$$\int_{B_R} \langle A(|Du|)Du, D\varphi \rangle \, dx = -\int_{B_R} f(u)\varphi \, dx.$$

To this end, let $\psi = \varphi\eta$, where for $0 < 2\varepsilon < R$,

$$
\eta(x) = \begin{cases} 0 & \text{for} \quad |x| \le \varepsilon, \\ 1 & \text{for} \quad |x| \ge 2\varepsilon, \end{cases}
$$

and such that $\eta \in C^1(\mathbb{R}^n)$, $0 \le \eta \le 1$ in $\mathbb{R}^n$, $|D\eta(x)| \le 2/\varepsilon$ for all x with $\varepsilon \le |x| \le 2\varepsilon$. Consequently, using ψ as a test function in $B_R \setminus \{0\}$, we get

$$
\int_{B_R \setminus B_{2\varepsilon}} \langle A(|Du|)Du, D\varphi \rangle \, dx + \int_{B_{2\varepsilon} \setminus B_\varepsilon} \langle A(|Du|)Du, \eta D\varphi + \varphi D\eta \rangle \, dx
$$

$$
= - \int_{B_R \setminus B_{2\varepsilon}} f(u)\varphi dx - \int_{B_{2\varepsilon} \setminus B_\varepsilon} f(u)\eta\varphi dx.
$$

Now

$$
\left| \int_{B_{2\varepsilon} \setminus B_\varepsilon} \langle A(|Du|)Du, \eta D\varphi + \varphi\eta \rangle \, dx \right|
$$

$$
\le \sup_{B_{2\varepsilon}} \left\{ \Phi(|Du|) \cdot \left[|D\varphi| + \frac{2}{\varepsilon}, |\varphi| \right] \right\} \cdot |B_{2\varepsilon}| = o(\varepsilon^{n-1})
$$

since $Du(0) = \mathbf{0}$ and Φ is continuous at $\varrho = 0$ by (A2). Moreover

$$
\left| \int_{B_{2\varepsilon} \setminus B_\varepsilon} f(u)\eta\varphi \, dx \right| \le \text{Const.}\,\varepsilon^n.
$$

Letting $\varepsilon \to 0$ we get the required conclusion.

Uniqueness of C^1 distribution solutions of (8.4.4). This is an immediate consequence of the comparison Theorem 2.4.1 and Proposition 2.4.2.　□

Proof of Theorem 8.4.3. That $m \mapsto u(r, m)$ is non-decreasing in the variable m follows from comparison, as above.

Continuity. Let $0 < m_1 < m_2$ and write $u_1(r) = u(r, m_1)$ and $u_2(r) = u(r, m_2)$. We claim that

$$
0 \le u_2(r) - u_1(r) \le m_2 - m_1, \qquad r \in [0, R]. \tag{8.4.15}
$$

Indeed by (8.4.9), for all $r \in [0, R]$,

$$
u_2(r) = m_2 - \int_r^R \Phi^{-1}\left(s^{1-n} \int_0^s t^{n-1} f(u_2(t)) dt \right) ds,
$$

$$
u_1(r) = m_1 - \int_r^R \Phi^{-1}\left(s^{1-n} \int_0^s t^{n-1} f(u_1(t)) dt \right) ds.
$$

Then by subtraction

$$u_2(r) - u_1(r) = m_2 - m_1 - \int_r^R \left[\Phi^{-1}\left(s^{1-n} \int_0^s t^{n-1} f(u_2(t))dt \right) \right.$$
$$\left. - \Phi^{-1}\left(s^{1-n} \int_0^s t^{n-1} f(u_1(t))dt \right) \right] ds.$$

The function Φ^{-1} is strictly increasing by (A2) and f is non-decreasing in $\mathbb{R}$ by (F2). Therefore, since $u_1 \le u_2$ in $[0, R]$ by monotonicity, one sees that the quantity in square brackets above is non-negative, and (8.4.15) is proved.

Proof that $u < m$ in B_R. By (8.4.9) it is enough to show that

$$I = \int_r^R \Phi^{-1}\left(s^{1-n} \int_0^s t^{n-1} f(u(t))dt \right) ds > 0 \quad \text{for } r \in [0, R).$$

Clearly $u > 0$ in some interval $(r_0, R]$ with $r_0 \ge 0$, and in turn $f(u(s)) > 0$ in $(r_0, R]$ by (F2). Therefore

$$I \ge \int_{\max\{r_0, r\}}^R \Phi^{-1}\left(s^{1-n} \int_{r_0}^s t^{n-1} f(u(t))dt \right) ds > 0,$$

as required.   □

Proof of Theorem 8.4.4

We begin with a preliminary result, of interest in itself.

Theorem 8.4.5. *If $u_1 = u(\cdot, m_1)$ has a dead core B_{S_1}, then $u_2 = u(\cdot, m_2)$, $m_2 < m_1$, has a dead core B_{S_2}, with $S_2 > S_1$. Similarly, if either $u_1(0) > 0$ or $u_1(0) = 0$ and $u_1(r) > 0$ for $r \in (0, R]$, then $u_2 > u_1$ in B_R when $m_2 > m_1$.*

Proof. To prove the first part of the lemma, assume for contradiction that $m_2 < m_1$, but either $u_2(r) > 0$ in $(0, R]$, or $0 < S_2 \le S_1$. In the first case the solutions u_1 and u_2 must cross at some point $r_0 \in (S_1, R)$. Then, applying Theorem 2.4.1 in B_{r_0} (always with the help of Proposition 2.4.2), we find that $u_1 \equiv u_2$ in $[0, r_0]$, which is an obvious contradiction since $u_2(r) > 0$ on $(0, r_0]$, while $u_2 \equiv u_1 \equiv 0$ in $[0, S_1]$. The next case $0 < S_2 < S_1$ leads to a contradiction in the same way, see Figure 8.5.

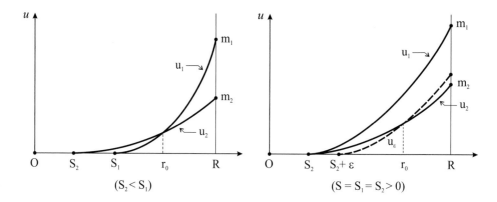

Figure 8.5.

The remaining case, when $S = S_2 = S_1 > 0$ needs more care. For $\varepsilon \in (0, R)$ define

$$u_\varepsilon(r) = \begin{cases} 0, & 0 \le r \le \varepsilon, \\ u_1(r - \varepsilon), & \varepsilon \le r \le R. \end{cases}$$

If $\varepsilon > 0$ is suitably small, then one has $m_1 > u_\varepsilon(R) > m_2 = u_2(R)$, while at the same time

$$u_2(S + \varepsilon) > 0 = u_1(S) = u_\varepsilon(S + \varepsilon). \tag{8.4.16}$$

Thus there is a point $r_0 \in (S + \varepsilon, R)$ where u_ε and u_2 cross, see the second case of Figure 5.

We assert that u_ε is a supersolution of (8.4.1) in the annulus $B_R \setminus \overline{B_\varepsilon}$. Indeed in this set we have

$$\text{div}\{A(|Du_\varepsilon|)Du_\varepsilon\} - f(u_\varepsilon) = \{A(|u_\varepsilon'|)u_\varepsilon'\}' + \frac{n-1}{r} A(|u_\varepsilon'|)u_\varepsilon' - f(u_\varepsilon)$$

$$= \left(\frac{n-1}{r} - \frac{n-1}{r - \varepsilon}\right)\Phi(u_\varepsilon') \tag{8.4.17}$$

$$= -\varepsilon \frac{n-1}{r(r - \varepsilon)}\Phi(u_1'(r - \varepsilon)) \le 0.$$

Observing that $u_2(0) = u_\varepsilon(0) = 0$, we can then apply the comparison Theorem 2.4.1 in B_{r_0}. Therefore $u_2 \le u_\varepsilon$ in $[0, r_0]$, which contradicts (8.4.16) for the specific value $r = S + \varepsilon$, and completes the first part of the proof.

To obtain the second part of the theorem, first assume for contradiction that $u_2(0) \leq u_1(0)$ when $m_2 > m_1$. Define

$$\tilde{u}_\varepsilon(r) = \begin{cases} u_2(0), & 0 \leq r \leq \varepsilon, \\ u_2(r - \varepsilon), & \varepsilon \leq r \leq R, \end{cases}$$

where ε is chosen so small that $m_2 > \tilde{u}_\varepsilon(R) > m_1 = u_1(R)$. Moreover $u_1(\varepsilon) > \tilde{u}_\varepsilon(\varepsilon)$, since by the final part of Theorem 8.4.2 we have $u_1'(r) > 0$ for $r \in (0, R]$. Hence there is a crossing point $r_0 \in (\varepsilon, R)$ where $u_1(r_0) = \tilde{u}_\varepsilon(r_0)$. As in (8.4.17) above, $\tilde{u}_\varepsilon$ is a supersolution of (8.4.1) in B_R. Thus $u_1 \leq \tilde{u}_\varepsilon$ in B_{r_0} by Theorem 2.4.1. In particular $u_1(\varepsilon) \leq \tilde{u}_\varepsilon(\varepsilon)$, which contradicts the fact that $u_1(\varepsilon) > \tilde{u}_\varepsilon(\varepsilon)$. Thus $u_2(0) > u_1(0)$.

That $u_2 > u_1$ in all B_R now follows at once, since otherwise u_2 and u_1 would cross at some value $r = r_0$ in which case comparison would lead to the absurd result $u_2 \equiv u_1$ in B_{r_0}. $\qquad \square$

Proof of Theorem 8.4.4. For the purpose of this proof, we suppose that there is some $m > 0$ for which $u(0, m) > 0$.

Existence of u_0. Define

$$m_0 = \inf\{m > 0 \; : \; u(0, m) > 0\}. \tag{8.4.18}$$

We claim first that $m_0 > 0$. Choose $\mu > 0$ so small that

$$R_{0,\mu} = \int_0^\mu \frac{ds}{H^{-1}(F(s)/n)} < R, \tag{8.4.19}$$

which of course is possible by assumption (8.4.2), see Lemma 4.1.2. Define $v(r) = w(r - S), r \in [S, R], S = R - C$, where w is the function constructed in the End Point Lemma 4.4.1, with $\sigma = 1/n$ and $C = R_{0,\mu}$.

We assert that v is a supersolution of (8.4.1) in the set $B_R \setminus \overline{B}_S$. In fact

$$\text{div}\{A(|Dv|)Dv\} = [\Phi(v')]' + \frac{n-1}{r}\Phi(v') \leq \left[1 + \frac{n-1}{r}(r - S)\right]\sigma f(v)$$

by (iii) and (iv) of Lemma 4.4.1. Thus

$$\text{div}\{A(|Dv|)Dv\} \leq \left(1 - \frac{n-1}{nr}S\right)f(v) \leq f(v),$$

as required. Then, since $v(S) = v'(S) = 0$, by defining v to be zero in B_S, the extended function v is a C^1 supersolution of (8.4.1) in all B_R, while also

$v(R) = \mu$. By the comparison Theorem 2.4.1 we find that $u(\cdot, \mu) \equiv 0$ in B_S. Therefore $m_0 \geq \mu > 0$ by (8.4.18) and the first part of Theorem 8.4.5. The assertion is proved.

Next, if (i) would be false, then $u_0(0) > 0$ and by Theorem 8.4.3 also $u(0, m) > 0$ for all values m sufficiently close to m_0, which would contradict (8.4.18). Property (ii) is again a direct consequence of the definition (8.4.18) of m_0 and Theorem 8.4.3. Finally if there is $m \in (0, m_0)$ such that the corresponding solution $u(\cdot, m)$ of (8.4.4) has no dead core, then $u(0, m) \geq 0$ and $u(r, m) > 0$ for $r \in (0, R]$. Thus by Theorem 8.4.5, with $m_1 = m$ and $m_2 = m_0$, we get $u_0(0) > u(0, m) \geq 0$, contradicting (i) and proving (iii).

Uniqueness of u_0. Suppose both m_0 and $\overline{m}_0$ have the properties (i)–(iii) of the theorem. Then $u_0(0) = u_0(0, m_0) = 0$ by (i), while $u(0, m) > 0$ when $m > \overline{m}_0$ by (ii). Hence $m_0 \leq \overline{m}_0$. Similarly $\overline{m}_0 \leq m_0$. Therefore $\overline{m}_0 = m_0$, as desired.

The case $m_0 = \infty$. If $u(0, m) = 0$ for all $m > 0$, then $u(\cdot, m)$ has a dead core for all $m > 0$. Otherwise there would be a value $\overline{m} > 0$ for which $u(0, \overline{m}) = 0$ and $u(r, \overline{m}) > 0$ for $r \in (0, R]$. Hence $u(0, m) > 0$ for $m > \overline{m}$ by Theorem 8.4.5, contradicting the assumption. This also justifies the earlier agreement that $m_0 = \infty$ in this case.   □

Remark. In summary, if m_0 is finite and $m > m_0$, then the solution $u = u(\cdot, m)$ of (8.4.4) is positive, namely $u(r, m) > 0$ for all $r \in [0, R]$. On the other hand, if $m < m_0 \leq \infty$, then the solution $u = u(\cdot, m)$ of (8.4.4) has a dead core $B_S \subset B_R$, $0 < S < R$.

The size of a dead core and proof of Theorem 8.4.1

Recall the assumption that $\Phi(\infty) = H(\infty) = \infty$, and let

$$R_0 = \int_0^\infty \frac{ds}{H^{-1}(F(s)/n)}. \tag{8.4.20}$$

Clearly $0 < R_0 \leq \infty$ since the integral is convergent at 0 by Lemma 4.1.2 with $\sigma = 1/n$. Of course the integral can possibly be divergent at ∞.

We prove two preliminary results.

Theorem 8.4.6. *We have*

$$m_0 = \infty \qquad if \quad R_0 < \infty \quad and \quad R \geq R_0, \tag{8.4.21}$$

while

$$m_0 \geq \overline{m} \qquad if \quad R < R_0, \tag{8.4.22}$$

where $\overline{m}$ is defined by the relation

$$R = \int_0^{\overline{m}} \frac{ds}{H^{-1}(F(s)/n)}.$$

Proof. The proof of (8.4.21) is essentially the same as the proof of the first part of Theorem 8.4.4, the only exception being that $C_{n,\mu}$ is replaced by R_0.

To obtain (8.4.22), we define $v(r) = w(r)$ as in the proof of Theorem 8.4.4 but with $S = 0$. Then by the End Point Lemma 4.4.1 there holds $v(0) = v'(0) = 0$, $v(R) = w(R) = \overline{m}$. Moreover v is a supersolution of (8.4.1). By virtue of Theorem 2.4.1, it follows that $0 \leq u(r, m) \leq v(r)$. Hence $u(0, \overline{m}) = 0$, and in turn from the definition (8.4.18) of m_0 we get $m_0 \geq \overline{m}$, as required in (8.4.22). $\qquad \square$

Theorem 8.4.7. *Let $m < m_0$, so that a dead core exists by Theorem 8.4.4-(iii). In particular the solution $u = u(\cdot, m)$ satisfies*

$$u \equiv 0 \qquad in \ B_S \subset B_R,$$

where

$$R - \int_0^m \frac{ds}{H^{-1}(F(s)/n)} < S < R.$$

If $R \geq R_0$, then for all $m > 0$ one has

$$R - R_0 < S < R.$$

Proof. The proof is the same as the first part of the proof of Theorem 8.4.4. $\qquad \square$

Remark. For any $\varepsilon > 0$, if m is suitably small (depending on ε), we have $R - \varepsilon < S < R$.

Proof of Theorem 8.4.1. *Part* (a). That $u \geq 0$ follows by Theorem 2.4.1 by comparing the given solution u with the trivial solution 0.

The constant function m is a supersolution of (8.4.1), so that again by Theorem 2.4.1 we have $u \leq m$ in Ω. In fact $u < m$ in Ω. To see this, let y be any point of Ω and B a ball in Ω centered at y. Let $v(\cdot, m)$ be the radial solution of (8.4.1) in B constructed in Theorem 8.4.2, with

$v(|x - y|, m) = m$ for $x \in \partial B$. Therefore $u(x) \leq m = v(|x - y|, m)$ for $x \in \partial B$, and in turn $u(x) \leq v(|x - y|, m) < m$ for $x \in B$ by the final part of Theorem 8.4.3.

Part (b). This is a direct consequence of Theorem 8.4.6.

Part (c). Let B be any ball compactly supported in Ω. Denoting the radius of B by $R - \varepsilon$, then by comparison, together with the remark after Theorem 8.4.7, we have $u \equiv 0$ in B when $m > 0$ is suitably small.

Since Ω' can be covered by a finite numbers of balls B, it follows that $u \equiv 0$ in Ω' when $m > 0$ is suitably small (depending only on the distance of Ω' to $\partial \Omega$).   □

The case $\Phi(\infty) < \infty$

This is the case, for example, for the mean curvature operator for which $\Phi(\infty) = H(\infty) = 1$. The proof of the principal Theorem 8.4.2 requires only the modification that the parameter m in (8.4.4) should be restricted so that

$$Rf(m) < n\Phi(\infty), \tag{8.4.23}$$

so that $\mathcal{T}$ in (8.4.9) is well defined. Moreover, for the application of the End Point Lemma 4.4.1 in the proof of Theorem 8.4.4 we also need the further restriction

$$F(m) < nH(\infty). \tag{8.4.24}$$

Denote by m_∞ the supremum of all $m > 0$ satisfying (8.4.23) and (8.4.24).

Then the main results stated in Section 8.4 remain true provided that the condition $m < m_\infty$ is assumed in all the statements. For instance we have the following analog of Theorem 8.4.1.

Theorem 8.4.8. *Assume the dead core condition* (8.4.2) *holds and let u be a solution of* (8.4.1), *with* $0 \leq u(x) \leq m$ *on* $\partial \Omega$ *for some positive constant* $m < m_\infty$. *Then the following properties are valid:*

(a) $0 \leq u < m$ *in* Ω.

(b) *Assume that* $R_0 = \displaystyle\int_0^{m_\infty} \dfrac{ds}{H^{-1}(F(s)/n)} < \infty$, *and let B_R be a ball with radius $R \geq R_0$, compactly contained in Ω. Then u has a dead core in Ω for all $m \in (0, m_\infty)$.*

(c) *If Ω' is any compactly contained set in Ω, then $u \equiv 0$ in Ω' provided that $m > 0$ is suitably small.*

It is not hard to show that if $\Phi(\infty) = \infty$, then necessarily $H(\infty) = \infty$, but it is possible to have $\Phi(\infty) < \infty$ and $H(\infty) = \infty$, as shown by $A(s) = 1/(1 + \sqrt{1 + s^2})$, with correspondingly

$$H(s) = \frac{1}{2}\left[\frac{s^2}{1 + \sqrt{1 + s^2}} - \log\frac{1 + \sqrt{1 + s^2}}{2}\right].$$

In this example $\Phi(\infty) = 1$, while $H(\infty) = \infty$.

The case $H(\infty) < \infty$ for unrestricted $m > 0$ was treated by Siegel in [102].

A dead core with bursts

It is known that when (1.1.7) holds and when $f = f(z)$ appropriately changes sign for $z > \delta$, there are non-negative radially symmetric solutions v of (8.4.1) having compact support; see for example [39]. Let R_* be the support radius of such a solution.

Next choose R and S in Theorem 8.4.7 so that $R_* < S < R$, and let w denote a corresponding dead core solution with small m. This being done, we can now replace the solution w on the set B_{R_*}, where it vanishes, by the solution v, thus obtaining a new solution u which is then positive in B_{R_*} and $B_R \setminus \overline{B_S}$, and otherwise vanishes. This solution may be considered as a dead core with a symmetric *burst* centered at the origin.

Of course, the same procedure may be repeated at other suitably chosen origins in B_S, giving rise to multiple bursts. Naturally a given ball B_S can accommodate only a certain number of bursts, but the larger are R and S, the more bursts which can be allowed (since possible values of the radius R_* are bounded away from zero). For details and further extensions the reader is referred to [82].

Notes

The results in Section 8.4 are taken from the paper [82]. In [82] the dead core problem for a weighted equation has been studied; for other related work we refer to the bibliography of this paper. A further related dead core theorem was given by Diaz and Véron [32].

Sperb [105] considers similar dead core problems for the special case of the Laplace operator, that is $A \equiv 1$. He estimates the critical value m_0 for general domains, but only for the homogeneous case $f(u) = \text{Const.} |u|^{q-1}u$, $0 < q < 1$. For balls his estimate is weaker than the exact result (8.4.8).

Theorem 8.4.4 for the general equation (8.4.1) seems to capture and extend many of the ideas of these earlier papers (for further extensions see [82]).

8.5 The strong maximum principle for Riemannian manifolds

Let $\mathcal{M}$ be an n-dimensional Riemannian manifold of class C^1, with controvariant metric tensor $[g^{ij}]$ continuous in local coordinates $x = (x^1, \ldots, x^n)$. Let u be a real-valued C^1 function defined on some open connected submanifold Ω of $\mathcal{M}$. The Riemannian norm of the gradient vector ∇u on Ω is then defined as the non-negative continuous function on Ω given in local coordinates by

$$|\nabla u|_g = \sqrt{g^{ij} \partial_{x^i} u \partial_{x^j} u}, \qquad \partial_{x^i} u = \frac{\partial u}{\partial x^i}.$$

Consider the variational integral

$$I[u] = \int_\Omega \{\mathscr{G}(|\nabla u|_g) + F(u)\} d\mathcal{M}.$$

The corresponding Euler–Lagrange equation is then

$$\text{div}_g \{A(|\nabla u|_g)\nabla u\} - f(u) = 0, \tag{8.5.1}$$

where div_g is the Riemannian divergence operator and $A(s) = \mathscr{G}'(s)/s$, $s > 0$, as in Section 1.1, see (1.1.3). More explicitly, in local coordinates $x = (x1, \ldots, x^n)$ in Ω, one has $d\mathcal{M} = \sqrt{g}\, dx$, where $g = 1/\det[g^{ij}]$. Then a direct calculation of the Euler–Lagrange equation yields

$$\frac{1}{\sqrt{g(x)}} \partial_{x^i}\left(\sqrt{g(x)} g^{ij}(x) A(|\nabla u|_g)\partial_{x^j} u\right) - f(u) = 0, \tag{8.5.2}$$

that is, exactly (8.5.1). When $A \equiv 1$ the differential operator in (8.5.2) reduces just to the manifold Laplacian, see [116, page 232].

A specific example is given by the variational integral

$$\int_\Omega \left\{\frac{1}{p}|\nabla u|_g^p + F(u)\right\} d\mathcal{M}, \qquad p > 1, \qquad \text{where} \quad d\mathcal{M} = \sqrt{g}\, dx \quad \text{on } \Omega,$$

introduced by Mossino ([64], page 40), though without the volume factor $\sqrt{g}$. Here of course $A(s) = s^{p-2}$, $p > 1$. Other examples are given also in [73], [77], [4].

Obviously (8.5.2) is the special case of

$$\partial_{x^i}\{a_{ij}(x, u)A(|Du|_g)\partial_{x^j}u\} - B(x, u, Du) \leq 0, \tag{8.5.3}$$

where $|Du|_g = \sqrt{g^{ij}(x, u)\partial_{x^i}u\partial_{x^j}u}$ is a gradient norm of Riemannian type and

$$a_{ij}(x, u) = \sqrt{g(x)}\, g^{ij}(x), \qquad B(x, u, \boldsymbol{\xi}) = \sqrt{g(x)}\, f(u).$$

With this motivation in hand in Section 9 of [81] we established a strong maximum principle for (8.5.3), but with a somewhat difficult proof. A strong maximum principle for the Riemannian equation (8.5.1)–(8.5.2), or for the corresponding inequality, can be treated more simply and under slightly lighter hypotheses. The result is as follows.

Theorem 8.5.1. *Let conditions* (A1), (A2), (F1) *and* (F2) *hold, as in Section 1.1. Assume that the Riemannian manifold $\mathcal{M}$ is of class C^3. Then the strong maximum principle is valid for the inequality*

$$\mathrm{div}_g\{A(|\nabla u|_g)\nabla u\} - f(u) \leq 0 \qquad in\ \Omega, \tag{8.5.4}$$

provided that $f(z) \equiv 0$ for $z \in [0, d]$, $d > 0$, or $f(z) > 0$ for $z \in (0, \delta)$ and (1.1.5) is satisfied.

Proof. In essence, we follow the proof of Theorem 1.1.1 in Section 5.1, but in the Hopf construction we replace the ball B_R tangent to the support of u by a small *geodesic ball* $\{x \in \Omega : s(x) \leq S\}$ centered at y and tangent to the singular set where $u = 0$, $Du = \mathbf{0}$; here $s(x)$ denotes the geodesic distance (with respect to the metric induced by the matrix $[g^{ij}]$) from the given center y to nearby points $x \in \Omega$. The existence of such a tangent ball can be shown exactly as in Hopf's original proof, at least provided that $|Ds|$ is equally bounded above and bounded away from zero.

To show this fact, we observe by Gauss' lemma (see [116], page 235) that

$$|Ds(x)|_g^2 = g^{ij}(x)\partial_{x^i}s(x)\partial_{x^j}s(x) = 1, \qquad x \neq x_0. \tag{8.5.5}$$

Thus, letting θ^2 and Θ^2 be the least and greatest eigenvalues of $[g^{ij}]$, we get

$$\Theta^{-1} \leq |Ds| \leq \theta^{-1},$$

as required.

Consider the geodesic annular set $G_S = \{x \in \Omega \ : \ S/2 \leq s(x) \leq S\}$ and let v be the unique solution of (4.2.1) given by Lemma 4.2.3, in k-dimensional space, where $R = S$ and the constant k will be determined later. In view of (1.1.5) of course $|Dv| > 0$ and so $|Dv|_g = |Dv|/|Ds| \geq \theta|Dv| > 0$.

Also by restricting the boundary value $v = m$ at $\partial B_{R/2}$ to be sufficiently small, one can maintain $\sup |Dv|_g \leq \Theta|Dv| \leq 1$.

The principal calculation, for $x \in G_S$, is the following:

$$\frac{1}{\sqrt{g(x)}} \partial_{x^i}\{\sqrt{g(x)}\, g^{ij}(x) A(|Dv|_g)\partial_{x^j} v\} - f(v)$$
$$= -\frac{1}{\sqrt{g(x)}} \partial_{x^i}\{\sqrt{g(x)}g^{ij}(x)\partial_{x^j} s\, A(w')w'\} - f(w)$$
$$= [\Phi(w')]' - \Delta s\, \Phi(w') - f(w) \geq [\Phi(w')]' - \frac{k}{s}\Phi(w') - f(w),$$

where k is an appropriate constant. The remaining part of the proof involves application of the comparison Theorem 2.4.1. To this end, we have to check (2.4.3) when $\hat{A}(x, \boldsymbol{\xi}) = \sqrt{g(x)}\, g^{ij}(x)A(|\boldsymbol{\xi}|_g)\boldsymbol{\xi}$, that is, in Riemannian notation,

$$\sqrt{g(x)}\langle A(|\boldsymbol{\eta}|_g)\boldsymbol{\eta} - A(|\boldsymbol{\xi}|_g)\boldsymbol{\xi}, \boldsymbol{\eta} - \boldsymbol{\xi}\rangle_{\mathscr{M}}$$
$$\geq \sqrt{g(x)}\big(\Phi(|\boldsymbol{\eta}|_g) - \Phi(|\boldsymbol{\xi}|_g)\big) \cdot \big(|\boldsymbol{\eta}|_g - |\boldsymbol{\xi}|_g\big)$$

since $\langle \boldsymbol{\xi}, \boldsymbol{\eta}\rangle_{\mathscr{M}} \leq |\boldsymbol{\xi}|_g|\boldsymbol{\eta}|_g$, and (2.4.3) now follows because Φ is strictly increasing by (A2). $\square$

The strong maximum principle Theorem 8.5.1 was given in [81]. For the corresponding necessity of the conditions in Theorem 8.5.1 we refer to [77].

For the Laplace equation there is of course no problem – all solutions which are $o(|x|)$ as $|x| \to \infty$, or are bounded either above or below, are constants.

Problems

8.1   The condition that f be non-decreasing in Theorem 8.1.1 can be weakened to the condition

$$\inf_{z \geq c} f(z) > 0 \quad \text{when } f(c) > 0$$
$$\sup_{z \leq c} f(z) < 0 \quad \text{when } f(c) < 0.$$

8.2 Show that for the Poisson equation (I) it is enough for the conclusions to hold that $u(x) = o(|x|^2)$ as $|x| \to \infty$.

8.3 Let the hypotheses of Theorem 8.1.1 hold.
(i) If u is a solution of (8.1.2) in an exterior domain, which is $o(|x|)$ as $|x| \to \infty$, show that $f(c) = 0$ for all values c which can be attained by the solution at ∞.

8.4 Carry out the details in the proof of Theorem 8.1.3.

[*Hint*: In proving (8.1.9) it is helpful to use Young's inequality.]

8.5 Prove Theorem 8.2.2 in the easier case in which $u > 0$ in Ω.

8.6 Prove Theorem 8.3.2 when $u \geq 0$ and the Neumann constant in (8.3.3) is positive and when $u \geq 0$ and $f(0,0) \geq 0$.

8.7 Prove that the solution $u = u(\cdot, m)$ of (8.4.4) must be of one of the following three types: (a) $u > 0$ in B_R; (b) $u(0, m) = 0$ and $u'(r, m) > 0$ when $r > 0$; (c) There exists $S \in (0, R)$ such that $u'(r, m) > 0$ when $S < r < R$ and $u \equiv 0$ in B_S.

8.8 Supply the details for the proof of Theorem 8.4.7.

8.9 Prove Theorem 8.4.8.

8.10 Show that if $\Phi(\infty) = \infty$ then necessarily $H(\infty) = \infty$.

Bibliography

[1] AFTALION, A. AND J. BUSCA, Radial symmetry of overdetermined boundary value problems in exterior domains, *Archive Rational Mech. Analysis*, **143** (1998), 195–206.

[2] ALEXANDROFF, A.D., A characterization property of the spheres, *Ann. Mat. Pura Appl.*, **58** (1962), 303–354.

[3] ALMGREN, F.J., A maximum principle for elliptic variational problems, *J. Funct. Anal.*, **4** (1969), 380–390.

[4] ANTONINI P., M. MUGNAI AND P. PUCCI, Singular elliptic inequalities on complete manifolds, *J. Math. Pures Appl.* **87** (2007), 582–600.

[5] BANDLE, C. AND I. STAKGOLD, The formation of the dead core in parabolic reaction-diffusion problems, *Trans. Amer. Math. Soc.*, **286** (1984), 275–293.

[6] BANDLE, C., R.P. SPERB AND I. STAKGOLD, Diffusion and reaction with monotone kinetics, *Nonlinear Analysis, T.M.A.*, **8** (1984), 321–333.

[7] BARDI, M. AND F. DA LIO, On the strong maximum principle for fully nonlinear degenerate elliptic equations, *Arch. Math.*, **73** (2000), 276–285.

[8] BARLES, G., G. DIAZ AND J.I. DIAZ, Uniqueness and continuum of foliated solutions for a quasilinear elliptic equation with a nonlipschitz nonlinearity, *Comm. Partial Diff. Equations*, **17** (1992), 1037–1050.

[9] BENCI V., D. FORTUNATO AND L. PISANI, Solitons like solutions of a Lorentz invariant equation in dimension 3, *Rev. Math. Phys.*, **10** (1998), 315–344.

[10] BENILAN, P., H. BREZIS AND M. CRANDALL, A semilinear equation in $L1(\mathbb{R}^n)$, *Ann. Scuola Norm. Sup. Pisa*, **4** (1975), 523–555.

[11] BERESTYCKI, H. AND L. NIRENBERG, On the method of moving planes and the sliding methods, *Bol. Soc. Brasil. Mat.*, **22** (1991), 1–37.

[12] BIDAUT-VÉRON, M.-F., Local and global behavior of solutions of quasilinear equations of Emden–Fowler type, *Arch. Rational Mech. Anal.*, **107** (1989), 293–324.

[13] BIDAUT-VÉRON, M.-F. AND S.I. POHOZAEV, Nonexistence results and estimates for some nonlinear elliptic problems, *J. Anal. Math.*, **84** (2001), 1–49.

[14] BIRINDELLI, I. AND F. DEMENGEL, First eigenvalue and maximum principle for fully nonlinear singular operators, *Adv. Diff. Equations*, **11** (2006), 91–119.

[15] BOMBIERI, E., E. DE GIORGI AND M. MIRANDA, Una maggiorazione a priori relativa alle ipersuperfici minimali non-parametriche, *Archive Rational Mech. Anal.*, **32** (1969), 255–267.

[16] BREZIS, H., Symmetry in Nonlinear PDE's, *Proc. Symposia in Pure Math.*, Amer. Math. Soc., **65** (1999), 1–12.

[17] CAFFARELLI, L.A. AND X. CABRÉ, Fully Nonlinear Elliptic Equations, *Amer. Math. Soc. Colloquium Publications*, **43**, 1995.

[18] CASTRO, A. AND R. SHIVAJI, Nonnegative solutions to a semilinear Dirichlet problem in a ball are positive and radially symmetric, *Commun. Partial Differ. Equations*, **14** (1989), 1091–1100.

[19] CELLINA, A., On the strong maximum principle, *Proc. Amer. Math. Soc.*, **130** (2001), 413–418.

[20] CONLEY C.H., P. PUCCI AND J. SERRIN, Elliptic equations and products of positive definite matrices, *Math. Nachrichten*, **278** (2005), 1490–1508.

[21] CORTÁZAR, C., M. ELGUETA AND P. FELMER, On a semilinear elliptic problem in $\mathbb{R}^n$ with a non-Lipschitzian nonlinearity, *Adv. in Diff. Equations*, **1** (1996), 199–218.

[22] COURANT, R. AND D. HILBERT, Methoden der Mathematischen Physik, I, II, Springer-Verlag, 1924, 1937, and Methods of Mathematical Physik, Vols. 1, 2, Wiley-Interscience, N.Y., 1962.

[23] CUESTA, M. AND P. TAKÁČ, A strong comparison principle for positive solutions of degenerate elliptic equations, *Diff. Int. Equations*, **13** (2000), 721–746.

[24] DAMASCELLI, L. AND F. PACELLA, Monotonicity and symmetry of solutions of p-Laplace equations, via the moving plane method, *Ann. Scuola Norm. Sup. Pisa Cl. Sci.*, **26** (1998), 689–707.

[25] DAMASCELLI, L. AND B. SCIUNZI, Regularity monotonicity and symmetry of positive solutions of m-Laplace equations, *J. Diff. Equations*, **206** (2004), 483–515.

[26] DANCER, E.N., Some notes on the method of moving plane, *Bull. Australian Math. Soc.*, **46** (1992), 425–434.

[27] DE GIORGI, E., Sulla differenziabilità e l'analiticità degli estremali degli integrali multipli regolari, *Mem. Accad. Sci. Torino, Cl. Sci. Fis. Mat. Nat., Ser. 3*, **3** (1957), 25–43.

[28] DIAZ, J.I., Nonlinear Partial Differential Equations and Free Boundaries, *Pitman Research Notes in Mathematics*, **106**, 1985.

[29] DIAZ, J.I., Some properties of solutions of degenerate second order PDE in non divergence form, *Appl. Anal.*, **20** (1985), 309–336.

[30] DIAZ, J.I. AND M.A. HERRERO, Estimates on the support of the solutions of some nonlinear elliptic and a parabolic problems, *Proc. Royal Soc. Edinburgh*, **89-A** (1981), 249–258.

[31] DIAZ, J.I., J.E. SAA AND U. THIEL, Sobre la ecuación de curvatura media prescrita y otras ecuaciones cuasilineales elípticas con soluciones anulándose localmente, *Revista Unión Matemática Argentina*, **35** (1990), 175–206.

[32] DIAZ, J.I. AND L. VERON, Local vanishing properties of solutions of elliptic and parabolic quasilinear equations, *Trans. Amer. Math. Soc.*, **290** (1985), 787–814.

[33] DOLBEAULT, J., P. FELMER AND R. MONNEAU, Symmetry and nonuniformly elliptic operators, *Differential Integral Equations*, **18** (2005), 141–154.

[34] EVANS, L.C., Partial Differential Equations, *Graduate Studies in Mathematics* **19**, Amer. Math. Soc. (1998).

[35] FARINA, A., Liouville-type Theorems for Elliptic Problems, in *Handbook of Differential Equations – Stationary Partial Differential Equations*, Ed. M. Chipot, Elsevier BV, **4** (2007), 60–116.

[36] FELMER, P. AND A. QUAAS, On the strong maximum principle for quasilinear elliptic equations and systems, *Adv. in Diff. Equations*, **7** (2002), 25–46.

[37] FLEMING, W.H. AND R. RISHEL, An integral formula for total variation, *Arch. Math.*, **11** (1960), 218–222.

[38] FRAENKEL, L.E., An introduction to maximum principles and symmetry in elliptic problems, *Cambridge Tracts in Mathematics*, Cambridge Univ. Press, 2000.

[39] FRANCHI, B., E. LANCONELLI AND J. SERRIN, Existence and uniqueness of nonnegative solutions of quasilinear equations in $\mathbb{R}^n$, *Adv. in Mathematics*, **118** (1996), 177–243.

[40] GIDAS, B., W.-M. NI AND L. NIRENBERG, Symmetry and related properties via the maximum principle, *Comm. Math. Phys.*, **68** (1979), 209–243.

[41] GIDAS, B. AND J. SPRUCK, Global and local behavior of positive solutions of nonlinear elliptic equations, *Comm. Pure Appl. Math.*, **34** (1981), 525–598.

[42] GILBARG, D., Some hydrodynamic applications of function theoretic properties of elliptic equations, *Math. Zeit.*, **72** (1959), 165–174.

[43] GILBARG, D. AND N. TRUDINGER, Elliptic Partial Differential Equations of Second Order, Springer-Verlag, 2nd edition, 1983.

[44] GRANLUND, S., Strong maximum principle for a quasi-linear equation with applications, *Ann. Acad. Sci. Fennicae, Ser. A, I. Mathematica Dissertationes Number* 21, (1978).

[45] HARNACK, A., Die Grundlagen der Theorie des logarithmischen Potentiales und der eindeutigen Potentialfunktion in der Ebene, *V.G. Teubner*, Leipzig, 1887.

[46] HOPF, E., Elementäre Bemerkungen über die Lösungen partieller Differentialgleichungen zweiter Ordnung vom elliptischen Typus, *Sitzungsberichte Preussische Akademie der Wissenschaften*, Berlin, 1927, 147–152.

[47] HOPF, E., A remark on linear elliptic differential equations of the second order, *Proc. Amer. Math. Soc.*, **3** (1952), 791–793.

[48] JOHN, F. AND L. NIRENBERG, On functions of bounded mean oscillation, *Comm. Pure Appl. Math.*, **14** (1961), 415–426.

[49] KAWHOL, B. AND N. KUTEV, Strong maximum principle for semicontinuous viscosity solutions to nonlinear partial differential equations, *Arch. Math.*, **70** (1998), 470–478.

[50] KELLER, J., On solutions of $\Delta u = f(u)$, *Comm. Pure Applied Math.*, **10** (1957), 503–510.

[51] KELLOGG, O.D., Foundations of Potential Theory, New York, Dover, 1954.

[52] KRYLOV, N.V., *Nonlinear Elliptic and Parabolic Equations of Second Order*, Reidel, Dordrecht, Netherlands, 1987.

[53] KRYLOV, N.V., On the general notion of fully nonlinear second order elliptic equations, *Trans. Amer. Math. Soc.*, **347** (1995), 857–895.

[54] LADYZHENSKAYA, O.A. AND N.N. URAL'TSEVA, Quasilinear elliptic equations and variational problems with several independent variables, *Uspehi Mat. Nauk*, **16** (1961), 19–90. English Translation in *Russian Math. Surveys*, **16** (1961), 17–92.

[55] LIEBERMAN, G.M., The natural generalization of the natural conditions of Ladyzhenskaya and Ural'tseva for elliptic equations, *Comm. Partial Diff. Eqs.*, **16** (1991), 311–361.

[56] LITTMAN, W., A strong maximum principle for weakly L-subharmonic functions, *J. Math. Mech.*, **8** (1959), 761–770.

[57] MAZ'YA, V.G., Some estimates for the solutions of second order elliptic equations, *Dokl. Akad. Nauk SSSR*, **137** (1961), 299–302; English Translation in *Soviet Math. Dokl.*, **2** (1961), 413–415.

[58] McNABB, A., Strong comparison theorems for elliptic equations of second order, *J. Math. Mech.*, **10** (1961), 431–440.

[59] MEYERS, N.G., An example of non-uniqueness in the theory of quasilinear elliptic equations of second order, *Archive Rational Mech. Anal.*, **14** (1963), 177–179.

[60] MITIDIERI, E. AND S.I. POHOZAEV, The absence of global positive solutions to quasilinear elliptic inequalities, *Dokl. Akad. Nauk*, **359**

(1998), 456–460; English translation in *Dokl. Math.*, **57** (1998), 250–253.

[61] MONTICELLI, D.D., Maximum Principles and Applications for a Class of Degenerate Elliptic Linear Operators, *PhD Thesis* at the University of Milan, Department of Mathematics "F. Enriques", a.y. 2005–2006.

[62] MOSER, J., A new proof of De Giorgi's theorem concerning the regularity problem for elliptic differential equations, *Comm. Pure Appl. Math.*, **13** (1960), 457–468.

[63] MOSER, J., On Harnack's theorem for elliptic differential equations, *Comm. Pure Appl. Math.*, **14** (1961), 577–591.

[64] MOSSINO, J., Inégalités isopérimétriques et applications en physique, Travaux en Cours, Hermann, Paris, 1984.

[65] NIRENBERG, L., A strong maximum principle for parabolic equations, *Comm. Pure Appl. Math.*, **6** (1953). 167–177.

[66] NIRENBERG, L., Estimates for elliptic equations in unbounded domains and applications to symmetry and monotonicity, in *Harmonic analysis and partial differential equations*, Lectures in Math., Univ. Chicago Press, 1999, 263–274.

[67] NITSCHE, J.C.C., Elementary proof of Bernstein's theorem on minimal surfaces, *Ann. Math.*, **66** (1957), 543–544.

[68] OLEINIK, O.A., On properties of some boundary problems for equations of elliptic type, *Math. Sbornik*, N.S. **30** (72), (1952), 695–702.

[69] OSSERMAN, R., On the inequality $\Delta u \geq f(u)$, *Pac. J. Math.*, **7** (1957), 1641–1647.

[70] PELETIER, L.A. AND J. SERRIN, Gradient bounds and Liouville theorems for quasilinear elliptic equations, *Ann. Scuola Norm Sup. Pisa Cl. Sci.*, **5** (1978), 65–104.

[71] PICONE, M., Maggiorazione degli integrali delle equazioni lineari ellittico-paraboliche alle derivate parziali del second'ordine, *Atti Rend. Accad. Naz. dei Lincei*, (6) **5** (1927), 138–143.

[72] PICONE, M., Maggiorazione degli integrali delle equazioni totalmente paraboliche alle derivate parziali del secondo ordine, *Ann. Mat. Pura Appl.*, **7** (1929), 145–192.

[73] PIGOLA, S., M. RIGOLI AND A.G. SETTI, Maximum principles and singular elliptic inequalities, *J. Funct. Anal.*, **193** (2002), 224–260.

[74] POLACIK, P., P. QUITTNER AND P. SOUPLET, Singularity and decay estimates in superlinear problems via Liouville-type theorems. Part I: Elliptic equations and systems, Duke Math. J., (2007), to appear.

[75] PORRETTA, A. AND L. VÉRON, Symmetry of large solutions of nonlinear elliptic equations in a ball, *J. Funct. Anal.*, **236** (2006), 581–591.

[76] PROTTER, M.H. AND H.F. WEINBERGER, Maximum Principles in Differential Equations, Englewood Cliffs, N.J., Prentice-Hall, 1967.

[77] PUCCI, P., M. RIGOLI AND J. SERRIN, Qualitative properties for solutions of singular elliptic inequalities on complete manifolds, *J. Differ. Equations*, **234** (2007), 507–543.

[78] PUCCI, P., B. SCIUNZI AND J. SERRIN, Partial and full symmetry of solutions of quasilinear elliptic equations, via the Comparison Principle, *Contemp. Math.*, special volume dedicated to H. Brézis (2007).

[79] PUCCI, P. AND J. SERRIN, A note on the strong maximum principle for elliptic differential inequalities, *J. Math. Pures Appl.*, **79** (2000), 57–71.

[80] PUCCI, P. AND J. SERRIN, The Harnack inequality in $\mathbb{R}^2$ for quasilinear elliptic equations, *J. d'Anal. Math.*, **85** (2001), 307–321.

[81] PUCCI, P. AND J. SERRIN, The Strong Maximum Principle Revisited, *J. Differ. Equations*, **196** (2004), 1–66; Erratum, *J. Differ. Equations*, **207** (2004), 226–227.

[82] PUCCI, P. AND J. SERRIN, Dead Cores and Bursts for Quasilinear Singular Elliptic Equations, *SIAM J. Math. Anal.*, **38** (2006), 259–278.

[83] PUCCI, P. AND J. SERRIN, Maximum Principles for Elliptic Partial Differential Equations, in *Handbook of Differential Equations – Stationary Partial Differential Equations*, Ed. M. Chipot, Elsevier BV, **4** (2007), 355–483.

[84] PUCCI, P., J. SERRIN AND H. ZOU, A strong maximum principle and a compact support principle for singular elliptic inequalities, *J. Math. Pures Appl.*, **78** (1999), 769–789.

[85] REDHEFFER, R.M., On the inequality $\Delta u \geq f(u, |\mathrm{grad}\, u|)$, *J. Math. Anal. Appl.*, **1** (1960), 277–299.

[86] REDHEFFER, R.M., Nonlinear differential inequalities and functions of compact support, *Trans. Amer. Math. Soc.*, **220** (1976), 133–157.

[87] REICHEL, W., Radial symmetry for elliptic boundary value problems on exterior domains, *Archive Rational Mech. Analysis*, **137** (1997), 381–394.

[88] RELLICH, F., Zur ersten Randwertaufgabe bei Monge–Ampèreschen Differentialgleichungen vom elliptischen Typus; differentialgeometrische Anwendungen, *Math. Ann.*, **107** (1932), 505–513.

[89] ROYDEN, H.L., Real Analysis, 2nd ed., London, Macmillian, 1970.

[90] SAFONOV, M.V., Harnack's inequality for elliptic equations and Hölder property of their solutions (Russian), *Zap. Nauchn. Sem. Leningrad Otdel Mat. Inst. Steklov*, **96** (1980), 272–287; (English) *J. Soviet. Math.*, **21** (1983), 851–863.

[91] SERRIN, J., On the Harnack inequality for linear elliptic equations, *J. d'Anal. Math.*, **4** (1954–56), 292–308.

[92] SERRIN, J., Local behavior of solutions of quasilinear elliptic equations, *Acta Math.*, **111** (1964), 247–302.

[93] SERRIN, J., The problem of Dirichlet for quasilinear elliptic differential equations with many independent variables, *Philos. Trans. Roy. Soc. London*, Ser. A, **264** (1969), 413–496.

[94] SERRIN, J., On the strong maximum principle for quasilinear second order differential inequalities, *J. Functional Anal.*, **5** (1970), 184–193.

[95] SERRIN, J., A symmetry problem in potential theory, *Arch. Rat. Mech. Anal.*, **43** (1971), 304–318.

[96] SERRIN, J., Entire solutions of nonlinear Poisson equations, *Proc. London Math. Soc.*, **24** (1972), 348–366.

[97] SERRIN, J., Commentary on the Hopf strong maximum principle, in *Selected Works of Eberhard Hopf with commentaries*, ed. by C.S. Morawetz, J.B. Serrin and Y.G. Sinai, *Amer. Math. Soc.*, Providence, 2002.

[98] SERRIN, J., A remark on the Morrey potential, in *Control Methods in PDE-dynamical systems*, Contemporary Mathematics **426** (2007).

[99] SERRIN, J., A maximum principle for mean-curvature type elliptic inequalities, *J. European Math. Soc.* **8** (2006), 389–398.

[100] SERRIN, J. AND H. ZOU, Symmetry of ground states of quasilinear elliptic equations, *Arch. Rat. Mech. Anal.*, **148** (1999), 265–290.

[101] SERRIN, J. AND H. ZOU, Cauchy–Liouville and universal boundedness theorems for quasilinear elliptic equations and inequalities, *Acta Math.*, **189** (2002), 79–142.

[102] SIEGEL, D., Height estimates for capillary surfaces, *Pac. J. Math.*, **88** (1980), 471–515.

[103] SIMON, J., Minimal varieties in Riemannian manifolds, *Ann. Math.*, **88** (1968), 62–105.

[104] SPERB, R., *Maximum Principles and their Applications*, Academic Press, New York, 1981.

[105] SPERB, R., Some complementary estimates in the dead core problem, *Nonlinear Problems in Applied Mathematics. In honor of Ivar Stakgold on his 70th birthday*, T.S. Angell, et al. (eds.), Philadelphia, (1996) 217–224.

[106] STAMPACCHIA, G., Problemi al contorno ellittici, con dati discontinui dotati di soluzioni Hölderiane, *Ann. Mat. Pura Appl.*, **51** (1960), 1–38.

[107] TALENTI, G., Best constant in Sobolev inequality, *Ann. Mat. Pura Appl.*, **110** (1976), 353–372.

[108] TOLKSDORF, P., On the Dirichlet problem for quasilinear elliptic equations in domains with conical boundary points, *Comm. Partial Differ. Equations*, **8** (1983), 773–817.

[109] TRUDINGER, N.S., On Harnack type inequalities and their applications to quasilinear elliptic equations, *Comm. Pure Appl. Math.*, **20** (1967), 721–747.

[110] TRUDINGER, N., Linear elliptic operators with measurable coefficients, *Ann. Sc. Norm. Sup. Pisa*, **27** (1973), 265–308.

[111] TRUDINGER, N.S., Harnack inequalities for nonuniformly elliptic divergence structure equations, *Inventiones Math.*, **64** (1981), 517–531.

[112] TURKINGTON, B., Height estimates for exterior problems of capillary type, *Pac. J. Math.*, **88** (1980), 517–540.

[113] VÁZQUEZ, J.-L., A strong maximum principle for some quasilinear elliptic equations, *Applied Mathematics and Optimization*, **12** (1984), 191–202.

[114] WALTER, W., Differential and Integral Inequalities, *Springer Verlag*, Berlin, 1964 (German), 1970 (English).

[115] WIDMAN, K., A quantitative form of the maximum principle for elliptic partial differential equations with coefficients in L^∞, *Comm. Pure Appl. Math.*, **21** (1968), 507–513.

[116] WILLMORE, T.J., An Introduction to Differential Geometry, Oxford University Press, Oxford,, 1993.

[117] ZIEMER, W., Weakly Differentiable Functions, *Graduate Texts in Math.*, **120** (1989), Springer-Verlag.

Subject Index

Author Index

Progress in Nonlinear Differential Equations and Their Applications (PNLDE)

Edited by
Haim Brezis, Université Pierre et Marie Curie, Paris, France and Rutgers University, New Brunswick, N.J., USA

Progress in Nonlinear Differential Equations and Their Applications is a book series that lies at the interface of pure and applied mathematics. Many differential equations are motivated by problems arising in diversified fields such as mechanics, physics, differential geometry, engineering, control theory, biology and economics. This series is open to both the theoretical and applied aspects, hopefully stimulating a fruitful interaction between the two sides. It will publish monographs, polished notes arising from lectures and seminars, graduate level texts, and proceedings of focused and refereed conferences.